THINKING ABOUT PROVINCIALISM IN THINKING

POZNAŃ STUDIES
IN THE PHILOSOPHY OF THE SCIENCES AND THE HUMANITIES

VOLUME 100

FOUNDING EDITOR

Leszek Nowak

(1943-2009)

EDITORS

Jerzy Brzeziński	**Joanna Odrowąż-Sypniewska**
Andrzej Klawiter	**Katarzyna Paprzycka** (editor-in-chief)
Krzysztof Łastowski	**Piotr Przybysz**
Izabella Nowakowa	**Mieszko Tałasiewicz**

ADVISORY COMMITTEE

Joseph Agassi (Tel-Aviv)	**Theo A.F. Kuipers** (Groningen)
Étienne Balibar (Irvine, CA)	**Witold Marciszewski** (Warszawa)
Wolfgang Balzer (München)	**Ilkka Niiniluoto** (Helsinki)
Mario Bunge (Montreal)	**Günter Patzig** (Göttingen)
Nancy Cartwright (London)	**Jacek Paśniczek** (Lublin)
Robert S. Cohen (Boston)	**Marian Przełęcki** (Warszawa)
Francesco Coniglione (Catania)	**Jan Such** (Poznań)
Dagfinn Føllesdal (Oslo, Stanford)	**Max Urchs** (Konstanz)
Jaakko Hintikka (Boston)	**Jan Woleński** (Kraków)
Jacek J. Jadacki (Warszawa)	**Ryszard Wójcicki** (Warszawa)
Jerzy Kmita	

Address: dr hab., prof. UW Katarzyna Paprzycka · University of Warsaw · Institute of Philosophy
Krakowskie Przedmieście 3 · 00-927 Warszawa · Poland · Fax ++48(0)22-826-5734
E-mail: kpaprzycka@uw.edu.pl

POZNAŃ STUDIES IN THE PHILOSOPHY OF THE SCIENCES AND THE HUMANITIES, VOLUME 100

THINKING ABOUT PROVINCIALISM IN THINKING

Edited by

Krzysztof Brzechczyn

&

Katarzyna Paprzycka

AMSTERDAM–NEW YORK, NY 2012

The paper on which this book is printed meets the requirements of
"ISO 9706:1994, Information and documentation - Paper for
documents - Requirements for permanence".

ISBN: 978-90-420-3630-7
E-Book ISBN: 978-94-012-0900-7
© Editions Rodopi B.V., Amsterdam – New York, NY 2012
Printed in The Netherlands

CONTENTS

IV. GLOBALIZATION, SOCIETY, CULTURE

INTRODUCTION

Science – in the broadest sense of the term, which includes the natural sciences, the social sciences and the humanities – is done in many places of the world. Yet there are only a few great centers of science. They attract funding and provide almost ideal opportunities for research and development in every respect. It is not surprising that that they also attract some of the best researchers. Some – but not all. For a variety of reasons, some of the best researchers, or ones who have that potential, may do science outside these centers, in the provinces. Aside of the normal problems encountered by every researcher, compounded by lesser financial and infrastructural support, they also need to overcome the social obstacles due to their place of work. Their ideas, hypotheses, conceptions are not treated in the same way as they would be if they were proposed in the center. Such problems are aggravated especially for the young researchers.

This leads to a cluster of problems and issues, some of which are tackled in more depth by the papers assembled in the volume. What is an intellectual province? Who are the provincial thinkers? What is the mark of provincialism? Do provincial (or central) thinkers have any special duties? Are there ways of overcoming one's own provincialism?

The volume is divided into four parts. The first part is comprised by some of Leszek Nowak's (1943-2009) writings on provincialism as well as on related issues. Leszek Nowak, the founder of the *Poznań Studies in the Philosophy of the Sciences and the Humanities* series, has been deeply aware of this division and its importance for the way that science is done. This volume includes the translations of three of his papers devoted to the topic.

While Nowak's thought is full of admiration for the creative powers of scientists, it is also filled with a deep recognition of the social chains that bind them. In "On the Hidden Unity of Social and Natural Sciences," Nowak argues that the social sciences and the natural sciences share a common methodological core. There are fewer methodological differences between them than is sometimes thought. Yet, it is also a fact that the progress attained in the natural sciences surpasses that achieved in the

In: Krzysztof Brzechczyn and Katarzyna Paprzycka (eds.), *Thinking about Provincialism in Thinking* (*Poznań Studies in the Philosophy of the Sciences and the Humanities*, vol. 100), pp. 7-11. Amsterdam/New York, NY: Rodopi, 2012.

social sciences. Nowak defends the hypothesis that what is responsible for this state of affairs are the social influences that affect the day to day work of the social scientists.

"The Structure of Provincial Thought" is one of Nowak's best known papers on provincialism despite the fact that it has been originally published in a provincial journal. The paper is addressed to a wider audience of non-specialists. Using the conceptual tools of the idealizational theory of science, he argues that a scientific school usually involves a division of labor between the creators (founders of the school), correctors and applicators. While, he argues, such a division is quite natural at the level of community, it leads to problems when the division is transferred onto communities themselves. Some communities become the creators' communities (centers), while others can at best aspire to being the correctors' or even only the applicators' communities (provinces).

In "Models of Scientific Research," Nowak addresses perhaps one of the most burning questions, viz. how to educate young researchers. He argues that the most common view of scientific research encourages the education mainly of applicators and, at best, of correctors. If Nowak is right about this then our system of educating young researchers has the added disadvantage of fortifying provincialism. He proposes an alternative way of educating researchers, who are to gradually acquire applicational, correctional as well as creative abilities. – Both of us treat this text as a required reading for our doctoral students.

The second and the third part of the volume complement each other. They comprise studies of the various conceptualizations of provincialism as well as discuss various examples of the relationship between provinces and centers, which span different disciplines, cultures, locations and times.

Jan Woleński discusses the role played by the division into centers and provinces in philosophy. He distinguishes two senses of the term 'national philosophy'. A philosophy can be national only incidentally – the term then denotes philosophy done by authors who share a national origin. A philosophy can be national in a stricter sense when there is a closer relation of the content of the philosophy to the nation. In particular, he considers the views of Kazimierz Twardowski, the founder of the Lvov-Warsaw school on this question. Woleński recalls Twardowski's polemic with Struve. Struve advocated the view that philosophy in Poland ought to become a truly Polish philosophy (in the stricter sense). Twardowski opposed the view and argued that it leads to isolationism, which is characteristic of a strand of provincialism. As Woleński convincingly shows, the polemic is as current today as it was a century ago.

Giacomo Borbone looks at Gramsci's conceptualization of Italian provincialism at the beginning of the last century. Borbone distinguishes two dimensions of provincialism. One is, once again, isolationism, the tendency to seclude oneself and to ignore external ideas. The other is the lack of originality. He shows that Gramsci tried to overcome in particular the isolationism of Italian culture. He argues that the attempt was not successful.

The second dimension of provincialism, the lack of originality, is explored in Mieszko Ciesielski's reconstruction of Witold Gombrowicz's thoughts on provincialism. Three types of attitudes that can be adopted by humans are distinguished. The author suggests that one is typical of the center while the others are typical of the provinces. An attitude characteristic of the reflective province provides not only a way to overcome provincialism but provides the tools to rise above a certain naivety of the center.

Adolfo García de la Sienra and Leandro Rodríguez Medina take up the question whether Hispanic-American philosophy exhibits provincialism. They attack the question by considering three different types of authors or attitudes that can be adopted: the integrated with the center whose place of residence happens to be Hispanic America, the "inserted" who, though not fully integrated, do try to contribute to the mainstream, and the self-excluded who pursue the path of isolationism. The authors provide an interesting account of the historical origins of these attitudes as well as of their current ramifications. They strongly emphasize that living in the province does not necessarily force one into provincialism.

Patryk Pleskot looks at the history of intellectual exchanges between French and Polish historians in the second half of the twentieth century. It is undeniable that the contacts resulted especially in methodological affinities. It is often also claimed that Polish historiography was influenced by these contacts. It is this claim that Pleskot subjects to scrutiny. He distinguishes inspiration from influence and argues that historians cannot hope to establish the latter. He also raises the more general question whether historical studies are, by the nature of their subject-matter, bound to be provincial.

Pleskot's question about the historical sciences has a mirror question about other sciences, taken up by Wenceslao Gonzalez. Gonzalez analyzes methodological universalism in science. He notes that the adoption of the theoretical view that all science shares a common methodology might lead to the practical attitude of methodological imperialism, according to which the only science worth pursuing and supporting falls into the universal strictures. Methodological imperialism contradicts the desideratum of scientific freedom. Moreover, Gonzalez argues that scientific universalism is ill-equipped to deal with the problem of complexity.

Eliza Karczyńska analyzes the different status of Orientalism in the West and in the East. It the West, it could be seen as an original theory, while, in the East, it is difficult to see as something other than an unreflective adoption of the Western perspective and as a sign of provincialism. Karczyńska points out how difficult it is to overcome provincialism for frequently such attempts are themselves provincial.

Cezary Kościelniak addresses the question of provincialism on many levels. His point of departure is a case study of a university that is provincial in a geographical sense. He asks whether such a university has special responsibilities to its region. He also discusses several factors that place the provincial universities at a disadvantage.

Krzysztof Brzechczyn provides a reconstruction of Nowak's reflections on the political thought of the Solidarność movement. Brzechczyn contrasts Nowak's diagnosis with quite contrary diagnoses of the thought by Timothy Garton Ash or Zdzisław Krasnodębski. What the latter take to be the virtues of that thought, Nowak takes to be the symptoms of provincialism. Brzechczyn distinguishes three dimensions of Nowak's critique. The eclectic political thought of Solidarność provided no clear theoretical vision (formal-internal deficiencies), it was incapable of offering a diagnosis of the situation (cognitive deficiencies) and it was incapable of indicating the appropriate course of action (policy deficiencies).

The papers assembled in the final part of the volume consider in depth (Urchs and Scheffler), in breadth (Przybylska-Czajkowska and Czajkowski), and in point (Paprzycka), the question of the impact of social processes, especially globalization, on the questions of provincialism in science and culture in general. Barbara Przybylska-Czajkowska and Waldemar Czajkowski consider the impact of the globalization and the development of a knowledge society on our culture at large. They look to Nowak's non-Marxian historical materialism and Wallerstein's world-systems theory for inspiration. Katarzyna Paprzycka offers a reconstruction of some of Nowak's ideas on provincialism. She argues that his model should be amended to take account of what she calls an intellectual superpower, for which there is no room in Nowak's initial model.

In their extremely rich and engaging paper, Max Urchs and Uwe Scheffler explore multiple factors that are responsible for the perpetuation of the province-center divide. They start with some historical roots of the scientific infrastructure. They explore the complex effects of the interplay between economic and political factors on the way that science is done. Finally, they consider the effects of a variety of social factors, in particular they consider the impact of the internet. They look at the relation between provinces in a geo-social sense, on the one hand, and provinces in a scientific sense, on the other. They take the latter to be characterized primarily by the incapacity to think on one's own account but also by a tendency to isolationism. They point out, however, that the negative flavor associated with the province in the geo-social sense is slowly giving way to a rather positive vision of tranquility. They speculate whether something like this process might not also take place in the scientific domain. Perhaps we will learn to cherish the thought of the provinces, of work conducted in peace, outside of the rat-race.

Let this last thought close our reflection on this topic for the present volume.

Katarzyna Paprzycka & Krzysztof Brzechczyn

PART I

LESZEK NOWAK ON PROVINCIALISM

Leszek Nowak (1943-2009)

ON THE HIDDEN UNITY
OF SOCIAL AND NATURAL SCIENCES (1998)[*]

ABSTRACT. The paper addresses the problem of the delay of the social sciences with respect to the natural sciences. It is argued that there are no special differences between them from a methodological point of view. The methodology of both can be understood in terms of the idealizational conception of science. Nor is the subject-matter the source of the problems. It is argued that it is the social placement of the social sciences within wider communities that is responsible for the delay.[**]

1. Problem. The Discussion of a Quick Solution

1.1. *Problem*

Here is "one of the most vexing intellectual problems of the present era" (Searle 1984, p. 71). The distinguished American philosopher asks: "Why do the methods of the natural sciences not given us the kind of payoff in the study of human behavior that they have in physics and chemistry?" (p. 71). He thus poses anew the question, which has been debated for a century: Are the human sciences sciences in the same sense of the word as the natural sciences? At first glance, the answer appears to be negative. Is consciousness not a basic human fact? Is consciousness not something quite unique especially when compared to the causal and mechanical world? A physical event causes another event and it itself is caused by yet another. None of them are constructs, pictures or representations of others – they are not *about* other events of that type. Yet mental events – our

---

[*] The paper appears in English translation for the first time. The Polish original "O ukrytej jedności nauk społecznych i przyrodniczych" appeared in *Nauka* 1 (1998), 11-42.
[**] The abstract has been added by the editors.

In: Krzysztof Brzechczyn and Katarzyna Paprzycka (eds.), *Thinking about Provincialism in Thinking (Poznań Studies in the Philosophy of the Sciences and the Humanities*, vol. 100), pp. 15-50. Amsterdam/New York, NY: Rodopi, 2012.

beliefs or suppositions – are capable of describing all other events (including themselves). They do so truly or falsely, but they are capable of remaining in semantic relations to the world, of being *about* it. In other words, they are capable of remaining in relations that cannot be grasped in natural terms. Is it not self-evident that a creature capable of all this stands out from the natural world, and that the science that studies this creature must be *sui generis*, based on different principles than those designed for the study of inanimate or at any rate asemantical world of things? And if so, then it would appear to follow that the criteria for the evaluation of progress in the natural sciences are useless when applied to the humanities. Should we not apply some other evaluative tools that would be specific for the human sciences? – This is the reasoning of many philosophers of science, including the author cited, who claims that "social sciences in general are about various aspects of intentionality" (p. 82) and that the source of their uniqueness is "the intrinsically mental character of social and psychological phenomena" (p. 84).

1.2. *The Irrelevance of Ontology to Theory*

The argument outlined above is based on two premises. One is arguably true: the phenomenon of consciousness is ontically unique. The other is the assumption that this fact has a methodological significance: the uniqueness of the ontology of a given discipline requires a unique methodology for the construction and evaluation of a theory in that discipline. Although the second assumption appears to be obvious, it is in fact highly questionable.

Mental phenomena are highly mysterious. But they are not alone. There are many phenomena explored by the most exact sciences that are mysterious for a philosopher. What are natural numbers? The most obvious answer that they are mental constructs does not survive an elementary criticism. If that were the case, natural numbers would be extended in time and facts about them would also have a temporal dimension. Yet it is absurd to say that $nm = mn$ at such and such a time because this is when a mathematician (which one?) has this thought. So to rescue this seemingly obvious answer, one has to appeal to the concept of a type of mental constructs, i.e. to their class. So, if push came to shove, one could hold that numbers are classes of certain mental constructs and that mathematical facts are classes of our judgments about relations among constructs. However, the problem is that classes are themselves mathematical objects and, on the most natural interpretation, they are ideal objects, which exist beyond space and time and so also beyond our minds. Would it not perhaps be better to conceive of all mathematical objects including numbers in this way from the start since their alleged reduction to the sphere of the mind (as well as to the sphere of signs) leads to the unwanted ideal entities anyway? But admitting ideas as full-blooded entities independent of us would lead to an incredible revision of our materialistic prejudices. It would undermine our ordinary ontology at its very foundation even before the notion of a physical thing appears, not to mention living things and conscious ones among them.

Even the ontology of elementary mathematics leads to quite mysterious puzzles. However, and this is the point I am after, this ontological puzzle of the arithmetic of natural numbers, which was posed already by Plato and which has not been solved to date, is irrelevant to the subject matter of arithmetics. First, isolated theorems about natural numbers were proved, they were then deductively systematized into theories, which were finally formally axiomatized and the notion of a natural number was precisely defined in terms of classes. All of this took centuries and was carried out in ignorance of the ontological nature of numbers. It transpires that the ontological uniqueness of natural numbers and lack of clarity as to their ontic status does not prevent the acquisition of exact knowledge about them.

It takes only a moment to realize that this is quite a general rule. Physical theories are about physical objects and they offer quite a lot of reliable knowledge about them. At the same time, however, it is highly unclear from a philosophical point of view what the nature of a physical object is. Indeed, even the fact of their existence is puzzling, of which one can convince oneself by examining the topic of the so-called ontological categories (cf. *References*). One does not need to argue that the same is true for time, space, causality and so on. There are ontological puzzles everywhere, yet their resolution turns out to be irrelevant to the acquiring of positive knowledge about areas they are about. The development of exact sciences shows that we can have a lot of knowledge, and exact knowledge at that, without understanding what this knowledge is about. This paradoxical methodological fact must be simply accepted like all other facts.

Leaving the explanation of this fact on a side (cf. *References*), we should, however, point out that it follows from what we have said above that human sciences are possible without philosophy. Indeed, the human science that is unquestionably successful is theoretical economics. It is capable of building explanatory theories, which sometimes employ quite sophisticated mathematics. It admits of quite precise quantitative measurements. And all of this has frequently quite efficient practical implications. Yet, even the most abstract economic models are about conscious beings, endowed with free will, capable of creating signs. The ontic uniqueness of humans does not interfere with the construction of these models. What of the fact that humans have free will if, confronted with the choice "to have more money or to have less money," they make the obvious choice with tiresome monotony (as witnessed by the market economy and the free-market economy mechanism). This fact is then modeled by the economists as the principle of income maximization, which is statistically confirmed to a great extent. It is clear that it is possible that it will not be confirmed since humans really do have free will and might want to have less. The point is that in general they do not want to want less and, leaving aside marginal cases, humans behave out of their own free will as if they decided not to use it. What of the fact that the very category *money* is about a certain type of signs, whose ontic status is very puzzling, as philosophers well know, if these puzzles have no bearing for the economists. That is why they have learned quite a lot of what they wanted to know about money. Thanks to their knowledge, governments can for, example, prevent hyperinflation and to some extent control inflationary processes. It seems that the ontic status of *money* has no more bearing on these issues than the ontic status of *numbers* has for mathematics.

Indeed, the supposition that the ontology of human beings determines the uniqueness of the human sciences is simply false. Fortunately so. Otherwise, science would have to wait for us, philosophers, to deal with our problems. If so, however, that it is not true that the ontic complexities of human beings are responsible for "one of the most vexing intellectual problems of the present era," to use Searle's phrase again. So, we face the problem of explaining why the social sciences are late in their development, the problem of the delay of social sciences.

2. The Social Sciences and the Method of the Natural Sciences

2.1. *The Presuppositions of the Problem of the Delay of Social Sciences*

This problem is usually posed with view to a certain methodological picture of the natural sciences. Research is supposed to begin with the observation of facts and with the search for regularities and repeatability of phenomena, which are captured in the laws of nature. These laws are generalizations derived from the observation of phenomena. Explanation of known facts and prediction of new ones consists in deducing what has happened or what will happen from certain scientific laws. This is what J. Searle (1984, pp. 71-72) says and he is not, by any means, alone. He then easily demonstrates that the humanities, which are concerned with conscious, rational and free beings, do not fit the pattern set by the natural sciences. Human beings interpret one another's deeds and ascribe aims to one another. Interpretation is what makes human research unique. Searle thus repeats Wilhelm Dilthey's view: we explain nature but understand culture. The procedure of interpretation extends beyond observation and its rules are, by no means, empirical generalizations.

All of this is true. The humanities do not fit the empiricist pattern of a science that registers facts, generalizes them into laws and derives new facts from these laws. However, this truth does not do very much to explain the sources of the delay of the humanities. For physics does not fit this patter either. The empiricist picture of the natural sciences, with which the humanities are often confronted, is false.

2.2. *Idealization in the Study of the Nature*

It suffices to take a look at just any physical law to notice that it is not a generalization of empirical facts. These laws are always equipped with assumptions that they hold for material points, inertial systems, ideally black bodies, ideal gases, etc. What we observe, on the other hand, are real entities: spatially extended bodies, non-inertial systems, non-ideal gases, etc. And these observations cannot be generalized to hold for ideal entities. Searle illustrates the notion of a natural generalization thus:

> [If], for example, we have a claim about a certain law describing the fall of bodies, and we know the initial position of a body, we can really deduce what will happen later. (p. 64)

The problem is that the law of free fall could not be an empirical generalization for a reason that was transparent to its author, who put it forward as holding in the conditions of ideal vacuum. For he understood that air resistance:

> disturbs all motion in an infinite variety of ways corresponding to the infinite variety of form, weight, and velocity of the bodies . . . hence, in order to handle this matter in a scientific way, it is necessary to cut loose from these difficulties; and having discovered and demonstrated the theorems, in the case of no resistance, to use them and apply them with such limitations as experience will teach.[*] (Galileo 1930, pp. 181-182)

Accordingly, Galileo's law of free fall has the form of a conditional:

---

[*] *Translator's note.* G. Galilei, *Dialogues Concerning Two New Sciences*, trans. H. Crew and A. de Salvio, internet edition: The University of Adelaide, http://ebooks.adelaide.edu.au/g/galileo/dialogues/complete.html. Access: September 7, 2012.

(G) If a body falls to the Earth, where air resistance is equal to 0, its initial velocity is also equal to 0, and gravitational acceleration g is constant, then the distance of the fall s equals $\frac{1}{2}gt^2$ (where t is the time of the fall).

The conditional is based on conditions that are violations of empirical facts (*idealizing assumptions*), of which the author is fully aware. For purely logical reasons, this type of claim (*idealizational statement*) cannot be applied to actually falling bodies. It can, however, be *concretized* into more realistic formulae that take into account some of the factors that have been disregarded. For example, by removing the assumption that the initial velocity is equal to 0, Galileo proposes the following concretization of the law of free fall (G):

(cG) If a body falls to the Earth, where air resistance is equal to 0, its initial velocity is greater than 0, $g = $ const., then $s = v_0t + \frac{1}{2}gt^2$.

Although this statement is closer to reality than the most abstract *idealizational law* (G), it cannot be applied to empirical facts without falling into contradiction. After all, it still assumes that the conditions of ideal vacuum obtain, so it is still an idealizational statement. It is only in experimental conditions where, for example, air resistance is "negligibly small" and the resulting disparities are tolerable for a given purpose, that it is possible to *approximate* the statement and obtain a statement with no idealizing assumptions (a *factual statement*):

(aG) If a body falls to the Earth, where air resistance is negligible, $v_0 > 0$, $g = $ const., then $s \approx v_0t + \frac{1}{2}gt^2$.

Galileo was partially wrong about the idealizing assumptions of the law of free fall (he thought that g is really a constant). Moreover, he did not realize just how many idealizing conditions there are (there are at least eleven in classical mechanics). Still the principle of the applicability of physical laws to empirical reality is quite clear already in this case: not generalization but idealization of empirical facts is its principle.

Numerous studies that analyze various disciplines and aspects of research in the natural sciences (cf. *References*) show the same: the essence of the scientific method used in the natural sciences is an intentional deformation of the empirical phenomena, which then strives to reconstruct them starting with an initial simplification. The construction of a theory consists in the proposal of the most basic, i.e. the most simplified, model, which encompasses a group of mutually related idealizational laws. The basic model reveals in a pure form a system of what are taken to be the basic regularities in observational phenomena. Subsequently, these assumptions are waved one by one and the initial model is extended (concretized) so as to capture more and more of the wealth of phenomena and so as to give an ever better approximation to registered facts. An *idealizational theory* is the sequence of such models beginning with the basic model. It is accepted if it explains many known facts in a given discipline to a sufficient degree of approximation.

It is usually the case that a currently accepted idealizational theory cannot explain all known facts. It is then corrected. Upon encountering a fact that is inconsistent with even the most concretized model of the theory, scientists try to identify the cause of the disparity, which is some factor that has not been as yet recognized in the theory. This factor will have to be idealized from. It is only on such an additional idealizing assumption that the claims of the basic model can then be accepted. As a result, the basic model becomes even more idealized. The factor will also have to be introduced to the theory. The new idealizing assumption will have to be waived and the claims of the model will have to be concretized by taking into account the impact of the new factor. The result of this process will be a new more adequate version of the initial theory. We will say that the new theory *corresponds dialectically* with the initial theory. Normal everyday theoretical activity in the natural sciences consists basically in the construction of theories that correspond with the initial one. One theory may correspond with the initial theory with respect to one additional idealizing assumption and with another theory with respect to a different idealizing assumption. Since the construction of such theories builds on others (scientists construct new theories that stand in relation of correspondence to theories, which correspond with the initial theory), normal science can be pictured as the construction of a tree, whose root is the initial theory. Such a tree could be called, to use Kuhn's term, a *paradigm*. The relation of dialectical correspondence with a given theory does not extend beyond the paradigm initiated by that theory.

The relation of *dialectical refutation*, on the other hand, does extend beyond the limits of a paradigm. All theories comprising a paradigm based on a given idealizational theory accept its essentialist view of a given domain. They accept as the principal factors (ones that cannot be omitted) in a given domain of phenomena those factors that were introduced in the basic model of the initial theory. They only supplement the collection of secondary factors (ones that can be omitted). A fundamentally new theory, which dialectically refutes the current paradigm introduces new hitherto unknown principal factors. In this sense, such a theory abandons the current way of viewing the "essence" of the investigated phenomena. However, it is a continuation of the old paradigm to some extent. For the theory takes into account some of the factors discovered in the old paradigm. After all the old paradigm had some explanatory power. When the theory that dialectically refutes the old paradigm is accepted, it once again turns out that it is far from the point of being able to explain all known facts with sufficient precision and so it must be corrected. In this way, a new paradigm, a new tree of theories that dialectically correspond with that theory, is constructed.

2.3. *The Methodological Pattern of Natural Sciences*

There are studies that show that not only classical but also contemporary physics fits something like this picture. Other studies show that the same is true for other natural sciences including such basic ones as the theory of evolution or genetics (cf. *References*). We may thus hold that the method of the natural sciences can be summarized in the following way.

Idealization. One counterfactually accepts idealizing assumptions that omit the influence of secondary factors on an investigated magnitude. One puts forward a simple formula that expresses the influence of principal factors on that magnitude. The formula holds under all and only those assumptions. Idealization consists in the formulation of an idealizational conditional. Its antecedent contains the full list of accepted idealizing assumptions while its consequent contains a formula, which specifies how the factors considered to be principal affect the investigated magnitude.

Concretization. The idealizational law is concretized by waiving subsequent idealizing assumptions. The formula of the idealizational law is thus modified by taking into account the influence of subsequent secondary factors. The process of concretization produces idealizational statements that are ever closer to observational data and whose formulae are ever more complex. In the simplest case, the product of the process will be a simple sequence of the idealizational law and its concretizations (idealizational structure).

Approximation. The process of concretization is carried out until one achieves a level of approximation that is sufficient for a given scientific discipline. In other words, if one were to waive all idealizing conditions remaining in the last concretization, then the disparity between empirical results and predicted results derived from the formula of that concretization would not exceed the threshold of disparity accepted in a given discipline.

Testing. The idealizational law is tested via concretization and approximation. In case the approximation of the last concretization in the idealizational structure is accepted to be false (the disparity between the predicted and empirical results exceeds the accepted threshold), the idealizational structure is not rejected. Rather, this is a point at which the process of concretization is restarted. The idealizational structure is rejected only if it turns out that none of the possible concretizations can reduce the disparities.

Correspondence. It can turn out that a given idealizational structure t is rejected because of the failure to match empirical data. This does not necessarily mean that the idealizational law with its initial formula is rejected. This formula can be saved by identifying new conditions that influence the investigated magnitude. In this way, a new idealizational structure t' is constructed. It begins with a new idealizational law. Its formula is the same as in the old idealizational law but it is based on additional idealizing assumptions, which abstract from the influence of the newly discovered factors. If the new structure t' is able to explain the disparities that were responsible for the rejection of t, it is then accepted.

Dialectical refutation. However, if the disparities are not explained by t' or any other structure t'', t''' . . . , which dialectically corresponds with t, then the formula of the initial idealizational law is called into question. The factors taken to be principal are no longer treated as immune from being disregarded. New factors are put forward to play this role. As a result, an idealizational structure s is constructed. Its idealizational law has a new formula. The law is subsequently concretized and finally approximated. The new idealizational structure s is accepted if it is able to explain the disparities that could not be explained by any structure in the family of structures initiated by t. One condition of its acceptance, however, is that the old formula enter as a component of some concretization of the new formula. If it turns out that the new structure does not explain some empirical data, a new family of structures s', s'', . . . , which correspond with s, is constructed.

Abstracting from many detailed questions discussed in the philosophy of science and adopting certain additional assumptions (however controversial, cf. *References*), one may accept that the above outline roughly depicts the methodological pattern of the natural sciences. At any rate, this hypothesis allows one to explain many additional facets of actual research practice: the role of empirical and thought experiments, the nature of the procedure of correcting empirical data on the basis of the theory of the measuring instrument, and so on, and so forth (cf. *References*). In sum, one can hold that the hypothetical model sketched above corresponds sufficiently closely to the research practice of the natural sciences. So the question we should raise is: What is the relation of the social sciences to this pattern?

2.4. *Idealization in the Study of the Human World: The Case of the Reproduction Theory*

To obtain an answer, we need to take a look at the research practice in the social sciences. We have already mentioned the models in theoretical economics. K. Marx's theory of reproduction is a classic example of modeling in economics. An economy modeled in this theory comprises the sector of production (sector 1) and that of consumption (sector 2). The global product P_1 of sector 1 is the sum of the constant capital C_1 of that sector, the variable capital (labor force) V_1 of that sector and the surplus value M_1 created in that sector. The same is true for second sector, where the respective symbols will be subscripted with '2'.

Marx begins his inquiry on the global process of reproduction with simple reproduction where the fund of accumulation is equal to zero:

> Simple reproduction, reproduction on the same scale, appears as an abstraction, inasmuch as on the one hand the absence of all accumulation or reproduction on an extended scale is a strange assumption in capitalist conditions, and on the other hand conditions of production do not remain exactly the same in different years (and this is assumed).[*] (Marx 1955, vol. II, p. 416)

Other simplifications concern value-relations:

> It is furthermore assumed that products are exchanged at their values and also that there is no revolution in the values of the component parts of productive capital. (Marx 1975, vol. II, p. 392; Marx 1955, vol. II, p. 414)

Furthermore,

> To the extent however that [the value-components of the total annual product] are partially and unevenly distributed, they represent disturbances which, in the *first* place, can be understood as such only as far as they are regarded as *divergences* from unchanged value-relations. (Marx 1955, vol. II, p. 415)

Next, the society under consideration consist only of two classes:

> Apart from this class, according to our assumption – the general and exclusive domination of capitalist production – there is no other class at all

---

[*] *Translator's note.* This and the following two quotations are taken from I. Lasker's translation of *Capital* (vol. II, ch. XX, §I). The internet edition: http://www.marxists.org/archive/marx/works/1885-c2/ch20_01.htm. Access: September 8, 2012.

except the working-class.[*] (Marx 1975, vol. II, p. 348; Marx 1955, vol. II, p. 365)

There is no other class in the model in question for, in the real world, there are many social classes and strata, which is also evident from the reading of Marx's historical studies (cf. *References*). Finally, the society in question is isolated from all external influences, which implies, among other things, that there is no trade exchange with other economies. And so on.

If we restrict ourselves to what has been said above, we obtain the following list of simplifying assumptions, which constitute the model of simple reproduction:

   (a) there is no accumulation ($A = 0$),
   (b) commodities are exchanged according to their values,
   (c) the values of commodities produced are constant,
   (d) the society is composed of two classes only: owners of the means of production and direct producers,
   (e) the society is fully isolated from other societies.

Because Marx himself extended his theory only by waiving assumption (a), we can present his model of simple reproduction in a somewhat simplified way:

(M1) If $A = 0$ then $P_1 = C_1 + V_1 + M_1$
(M2) If $A = 0$ then $P_2 = C_2 + V_2 + M_2$

From these schemata, Marx derives the balance condition for the model of simple reproduction. It turns out that the simplified economy in question is in balance as long as constant capital in the consumption sector is equal to the sum of variable capital and surplus value of the production sector: $C_2 = V_1 + M_1$.

In Chapter 21 of volume II of *Capital* Marx waives idealizing condition (a) and considers an economy that accumulates, though it still satisfies the remaining idealizing assumptions. The accumulation fund is used to enlarge the means of production (M^C) and to hire more labor force (M^V). In the current more realistic model of enlarged reproduction, $A = M^C + M^V > 0$. Schemata (M1)-(M2) are modified accordingly:

(cM1) If $A > 0$ then $P_1 = C_1 + V_1 + M_1 + M_1^C + M_1^V$
(cM2) If $A > 0$ then $P_2 = C_2 + V_2 + M_2 + M_2^C + M_2^V$

---

[*] *Translator's note.* The quotation is taken from I. Lasker's translation of *Capital* (vol. II, ch. XVII, §II), the internet edition, *op. cit.*

As Marx shows, the balance condition for this somewhat more realistic model is: $C_2 + M_2^C = V_1 + M_1 + M_1^V$.

This theory was still very far from the complexities of empirical economies. No wonder that many authors have been finding additional reasons why the schemata (cM1)-(cM2) deviate from reality. One of them was the assumption implicitly taken by Marx, which was not very natural, viz. that all economic sectors can be viewed as producing either the means of production or consumption. There seems to be no place for the production of arms, for example. This problem was posed by the Japanese theoretician Nonomura, who showed that, from an economic point of view, the production of arms must be treated as a separate sector. The production of arms is funded by the state from taxes, which Nonomura takes to be part of the surplus product. Marx's models are too poor to capture the role played by the state – there is no place for it in the models of simple and enlarged reproduction. Marx's list of assumptions must thus be enriched with the following idealizing condition:

(f) taxes T in a given economy are equal to zero.

The formulae of simple reproduction known from *Capital* obtain given assumptions (a)-(f). Marx's claims (M1)-(M2) must thus be reformulated for simple reproduction, where the production of arms is abstracted from:

(N1)   If $A = 0$ & $T = 0$ then $P_1 = C_1 + V_1 + M_1$

(N2)   If $A = 0$ & $T = 0$ then $P_2 = C_2 + V_2 + M_2$

The modification of these schemata consists in the waiving of the newly introduced assumption (f) and in the appropriate correction of the formulae. Nonomura assumes that part of T is supplied to the first sector while the other part is spent as wages and, consequently, supplied to the second sector. Thus:

(cN1) If $A = 0$ & $T > 0$ then $P_1 = C_1 + V_1 + M_1 + T_1$

(cN2) If $A = 0$ & $T > 0$ then $P_2 = C_2 + V_2 + M_2 + T_2$

The formula for the third sector is:

(N3)   If $A = 0$ & $T > 0$ then $P_3 = C_3 + V_3 + M_3 + T_3$

Finally, Nonomura repeats Marx's concretization and obtains the following schemata of enlarged reproduction taking into account armaments:

(ccN1) If $A > 0$ & $T > 0$ then $P_1 = C_1 + V_1 + M_1 + M_1^C + M_1^V + T_1$

(ccN2) If $A > 0$ & $T > 0$ then $P_2 = C_2 + V_2 + M_2 + M_2^C + M_2^V + T_2$

(cN3) If $A > 0$ & $T > 0$ then $P_3 = C_3 + V_3 + M_3 + M_3^C + M_3^V + T_3$

He also formulates somewhat more complex balance conditions for the above models.

Nonomura's corrections were quite simple. That is why they serve as a good illustration of theoretical work in economics. The reproduction schemata were modified in the way described above many times. Lange's model takes into account the allocation of accumulation in sectors other than those where it is produced, i.e. it incorporates the transfer of accumulation across economic sectors. Kawakami's model takes into account the service sector. Sweeze-Lange's model disaggregates the consumption sector. Nagels has put forward a model of an open economy. And so on. All these theories stand in the same methodological relation to Marx's original theory as Nonomura's theory. And the very same thing happens in other orientations in theoretical economics. Further illustrations will need to be sought in *References*. Here I will restrict myself to stating one obvious conclusion.

It is clear now that the method used in the construction of reproduction schemata is almost exactly the same as the methods used in physics. In the same way, one forgets about the complexities of this world and escapes into idealized worlds in order to grasp simple regularities. In the same way, one then complicates these regularities by coming closer and closer to empirical facts. The science of economics, which deals with conscious, free, sign creating beings uses the methodological pattern of natural sciences nevertheless. To be on the safe side, let us now turn to consider very roughly whether the use of the method of idealization is rare or typical in the humanities (for a more extensive argumentation, cf. *References*).

2.5. *Idealization in the Study of the Human World: A Survey*

Linguistics is another discipline besides economics that celebrates great triumphs in recent decades. Noam Chomsky is the author of a real theoretical breakthrough. His linguistic theory is quite intentionally constructed as an idealizational theory. According to Chomsky (1966, p. 37), in normal conditions speech undergoes many (some violent) deformations, which in and of themselves do not tell us very much about the underlying linguistic schemata. Thus:

> The linguistic theory is concerned primarily with an ideal speaker-listener, in a completely homogeneous speech-community, who knows its language perfectly and is unaffected by such grammatically irrelevant conditions as memory limitations, distractions, shifts of attention and interest, and error (random or characteristic) in applying his knowledge of the language in actual performance (Chomsky 1965, p. 3)

Chomsky's theory is a reconstruction of the linguistic competence of this ideal speaker-listener. It is composed of three models. According to the first idealized model of the grammar of component structure, the speaker-listener constructs his sentences only out of simpler components and he does not impose any phonetic shape on them. Under these assumptions, the linguistic competence of a speaker-listener comprises generative rules that assign to each sentence a structural description, which determines the elements of which the sentence is composed as well as their order and organization. Many typical sentences of a natural language can be generated in this way but not all. One needs to wave the first assumption and consider the phenomenon of the transformation of some sentences into others. This is done in the second model, which presents a more complete set of rules of linguistic competence, the rules of generative-transformational grammar. Neither generative nor generative-transformational grammar explains how the syntactic structures take the form of speech. This is the task of the third model, yet closer to reality, the model of phonological grammar, which assigns phonetic shape to the constructs of generative-transformational grammar.

It is clear even from this cursory description that not only the author's intention but also the methodological structure of Chomsky's theory satisfy the typical demands of the method of idealization and concretization well respected in the natural sciences. Also in this case, the method employed in the humanities is the same as that in the natural sciences. But not only in this case. We will remember that the most typical example of the method employed in the humanities that is taken to be different from that employed in the natural sciences is interpretation, whose point is to uncover the intention of the author of some work, e.g. some text. Indeed, claims based on the understanding of symbols are supposed to be the most basic and unique type of claims in the humanities. Let us take a look at the nature of this work by considering the thought process of a lawyer. We will do so very cursorily, letting the reader find more precise reconstructions in the *References*.

Legal interpretation consists in the reconstruction of unambiguous norms, i.e. expression of the form: persons with properties W in circumstances O ought to realize (or refrain from realizing) the state of affairs p. Legal interpretation is thus a reconstruction of norms $n_1, \ldots, n_m$, $\ldots$ from a legal text. The point is that one accepts certain assumptions about the author of this text when one performs such a reconstruction. We will easily see that they are idealizing in nature. And so, one assumes that the legal text does not contain inconsistencies. An interpretation would be unacceptable were it to reconstruct a norm as demanding of persons W in circumstances O to realize p and as forbidding those same persons in those same circumstances to realize p. One assumes that the legal text is logically closed. If it generates norm n then it generates every logical consequence of *n*. One ascribes to the author of a legal text a complete and justified empirical knowledge. And so on. The legislator is viewed by lawyers as a logically consistent and competent person, who is also fully competent in any domain that is subject of the regulation. Who is such a legislator? It is not the parliament, which is what it is, as everyone can see and as lawyers see most acutely. The "legislator," conceived of as the author of the legal text, is an idealizational construct. Its relation to real parliaments is the same as the relation of the material point to real bodies. The so-called legal logic (traditional priniciples *lex posterior derogate priori*, *lex specialis derogate generalis*, etc.) implicit in the interpretational work of lawyers is a complex idealizational construction based on multiple models, like those used in physical or economic theories. The initial rules of this logic are analogous to the basic idealizational laws. Subsequent rules take into account more and more circumstances, which determine the direction of interpretation. In those further rules, the excessive postulates of the ideal legislator are made more realistic. Thus these rules constitute a kind of concretization of the initial ones. There are no rules, however, not even in the most complex systems, that would waive the assumption of rationality (the maximization of preferences). The requirement of rationality is never waived in any interpretation of the law. Rightly so for the reduction of the law to the level of rationality of some real members of parliament could lead to catastrophic results.

2.6. *Idealization as a Universal Method*

It is clear from the cursory survey of the methods used in various scientific disciplines that, in principle, i.e. with a tolerance for some unique features (such as the legal prohibition to "concretize beneath the level of the rationality of the legislator"), they consist in the realization of the same methodological requirements. These are: to simplify the object of study, to propose hypothetical regularities for a simplified system, to gradually modify and complicate them so that they reflect empirical regularities. These requirements (of abstract reconstruction rather than generalization of the observations of the real world) are themselves related to so abstract properties of cognitive procedures that these procedures remain the same independently of whether they are applied to elementary particles, biological populations, entrepreneurs, or communicators.

The conclusion of the second part of our argument is this: if we confront the social sciences not with the way that the humanists conceive of the natural sciences but with the natural sciences themselves, then we will see that these sciences use basically the same cognitive method as is used by the natural sciences. Any differences here are the differences of aims, conditions or contexts of use of basically the same method. At any rate, these differences cannot be attributed to the differences between these two groups of sciences. For example, economics is closer to physics than is physics to paleozoology. The latter is closer to historiography than to economics. If the assumptions on which our argument is based were really true then the view that the social sciences are separated from the natural sciences by the methodological Chinese Wall would be a myth.

It would be natural to pose the following question: If things look really so good for the social sciences, then why do they look so bad? We thus return to the problem of the delay of the social sciences. Before we turn to this question, we should devote some time to clarify a certain misunderstanding, perhaps not too deep but quite bothersome for science or for its good name at any rate.

3. Science and Human Spirituality

3.1. *A Certain Stereotype*

All that was said so far could be summarized by means of the following philosophical hypothesis: to understand some system one must first idealize it and then gradually reconstruct it by means of concretization. It looks as if it does not matter very much if the system in question is natural or social, or if it comprises unconscious things or conscious persons. Does this not mean that we are ignoring the uniqueness of human spirituality? What about the distinctness of human existence?

This fear seems to be based on a deeply rooted judgment (or perhaps prejudice), which locates science beyond the sphere of spiritual culture. On this view, spiritual matters are to be split between religion, philosophy, literature, painting, etc., while science is but a more subtle craft . . . One sometimes even gets the impression that some artists or priests begin to look down at "shallow science." This is arguably due to the cultural atmosphere of our times, which is skeptical or even suspicious of science. This atmosphere is itself based on an empiricist philosophy, which fails to understand what scientific work really consists in. I am going to give some arguments in support of this supposition, of course.

3.2. *Science and Literature*

According to the wide-spread stereotype, upheld even by those great minds that broke many stereotypes such as Bolesław Leśmian or Witold Gombrowicz, literary and scientific work are separated by an unbridgeable gulf. Science deals with the observation of facts and their generalization onto charts, while literature is a fine sphere, it is a matter of imagination expressed in the construction of fictional worlds. Such a picture of science has nothing to do with reality. We have seen that fictitious worlds are constructed in science with as much zeal as in literature. Although Józef K. is an ontically different object than Józef Kowalski, he is quite close to the material point from an ontological point of view. After all, we have seen that the most basic model of every scientific theory, whether natural or social, is about idealized worlds. It is only in further more realistic models that the theory returns to our world to approximate it. This tendency to return from ideal worlds to our world (the method of concretization) is indeed wheat distinguishes it from literature. But the point of departure is very similar.

The justification for the "construction of fictitious worlds" is the same in both areas. The point is to discover essential components of reality that are hidden and blurred in the complexities of empirical facts. Both science and literature connote in their "fictions" what is essential in reality they are about.

> In developing murderous plots, the authors of Greek tragedy have not dealt with the psychology of killers. They have spoken about *all* people. And they are great to the extent that the audience shares the truth of their judgments. Similarly, in showing emotionally deviant characters, Dostoyevski was not speaking about the mentally ill but was revealing the motives of ordinary people. He remains a literary genius to the extent that his readers can find those motives among their own (Falkiewicz 1989, p. 22, original emphasis).

However, literary writers stop at the construction of those worlds. They leave the intuitive corrections at the hands of the literary critics and of their readers with their everyday experience at their disposal. Scientists, on the other hand, systematically traverse the distance from their "fictions" to our world. They systematically confront the predicted results with independently gathered empirical data. Clearly there is a difference: literature is *not* science. But it is also not something dramatically different from science either! It is this kinship of the literary and the scientific approach to reality that is frequently not understood in the literary community. It is likewise not understood by scientistically minded philosophers of science who identify science not with the modeling of the world but with the empirical content of the models. The former and the latter see only the end point of science without understanding the starting point. But it is the starting point, which is much more significant and which is evidence for a far-reaching kinship of science and literature. For the (logical) starting point is absolutely crucial in all thinking processes. Science and art have a very close starting point, as we shall see.

3.3. *On the Structure of Imagination*

The same holds for other domains of spiritual culture. As long as one believes that science does not extend beyond the empirical world, it must appear to be something fundamentally distinct from religion. There is nothing more distinct from an empirical body that leaves traces in our senses than God. But what if one understands that the construction of ideal entities is part and parcel of the essence of science? An omniscient being knows all and only truths: x is an omniscient being just in case for every p, x believes that p iff p is a fact. Is this not something like an idealization? Perhaps despite many differences, the inertial system or ideally homogenous cosmos etc. are closer to mythical than to empirical entities. This topic has been only initially explored but it is so important, also from a more general point of view, that we will take a brief look at it (for a fuller discussion, cf. *References*).

Idealization is but one of a group of procedures, which could be called *deformational* for want of a better name. They consist in a counterfactual (contrary to our knowledge) deformation of an object. Let us looks at these procedures more systematically. Let an empirical object e and a space of mutually independent properties $A, \ldots, Z$ be given. Let us suppose that e has some of those properties $A, B, \ldots, K$ but lacks others: $-L, -M, \ldots$, $-Z$. Let each of those properties pertain to e in a certain degree. We will also assume that each of the properties has a minimal degree as well as a maximal degree. This completes our assumptions. We can distinguish four types of basic deformational procedures.

The procedure of *potentialization* consists in counterfactually taking the object e to have one of its properties in a degree that is different from the actual one. If the value attributed to e is greater than actual, we will talk about *positive potentialization*. We will speak of *negative potentialization* if the attributed value is smaller than the actual one. When I imagine myself in a couple of years, the white-haired man I have before my eyes is the result of positive potentialization performed on my own graying self. He is my positive potentiale with respect to hair color. The potentialization of an object e with respect to property A (where the actual value of A for e is a_i) consists in the counterfactual postulation of an object e', which has A but in a different degree a_j. In a limiting case of positive potentialization, we postulate an object that has a given property in the maximal degree; such a procedure could be called *mythization*. Analogically, the postulation of an object that has a given property in the minimal degree is a limiting case of negative potentialization; such a procedure could be called *ideation*. Potentialization is a "quantitative" form of deformation, so to speak. What changes is only the intensity of the properties that the object has. The set of properties that an object has and lacks remains the same, so the "qualitative" equipment of the object is untouched in the case of potentialization. We will say that potentialization is a weakly deformational procedure.

Reduction, by contrast, is a strongly deformational procedure. It consists in impoverishing the object by counterfactually subtracting some of its properties. In the above terms, reduction with respect to a property K consists in the postulation of an object e', which will lack this property rather than having it. The reduct e' of e has the same properties in the same degrees but, in addition, it fulfills the counterfactual condition that assigns to it $-K$ in place of K.

Transcendentalization is another strongly deformational procedure. It is the reverse of reduction. It consists in the counterfactual enrichment of an object by means of additional properties. The lack of a certain property in the original is replaced with such a property. The transcendens of an object e with respect to L is an object e' with the same properties in the same degrees as e, which in addition has property L, which e actually lacks (has $-L$).

All these procedures are counter-empirical. They do not reproduce the state of the empirical world but change it. They deform the property equipment of empirical objects, whether with respect to degrees or the properties themselves. They postulate their counterparts which share at most some of the property equipment of the originals. The limiting case of the reducts of an empirical object is an empty object, deprived of all properties (equipped only with their lacks). The limiting transcendens is the complete object, which is deprived of any lacks but equipped with all properties from a given space of properties. Human imagination not only is not limited to the empirical world but necessarily extends beyond it. One could say somewhat metaphorically in this context that human imagination in principle explores (various types of) other worlds.

3.4. *Science as the Expression of Human Spirituality*

Scientific imagination is no exception here. Indeed, we can capture all fruits of human spirit in these terms, including art, religion as well as science. *Idealization* can be identified as the combination of reduction and negative potentialization or ideation in the limiting case. Indeed, when one assumes that the Earth is a material point, one does two things at the same time. First, one reduces all the properties of the Earth except for mechanical ones. Second, one takes some of those remaining properties (e.g. dimensions) to be equal to zero, i.e. one performs ideation. The object under consideration ("theoretical Earth") is thus a reduct and a limiting negative potentiale of the planet we inhabit. Idealization is typical for science, as we have seen.

The combination of reduction and positive potentialization (mythization in the limiting case) can be called *fictionalization*. This procedure is characteristic for art. A literary character lacks some of the properties of the people that it portrays ("Life is too boring to replicate in literary action," as writers say), but has other properties to an exaggerated extent. In other words, it is subjected to reduction and positive potentialization. The figure portrayed in a painting is by definition a reduct of the model (some of whose properties cannot be visually represented), while those properties that are represented are emphasized with respect to the features empirically exhibited by the model. A mime again is by definition a reduct (he does not use speech); yet the properties that he does represent, he exhibits with a force that we do not encounter in everyday life. And so on. Fictions are not ideal objects but they are indeed something akin to them, as we have suggested above.

Finally, *absolutization* is the combination of transcendentalization and positive potentialization (mythization in the limiting case). Indeed, God is an absolute. It is a complete object (exhibiting all properties), each in the highest degree: God is all-knowing and knows everything with maximal certainty, he loves all creation and infinitely so (in the highest possible degree) and so on. Religion then, and the sphere of myth more generally, is the domain of absolutization.

We see then that the human spirituality expresses itself in the imagination, i.e. in the ability to deform the world in thought. This ability is used by science as it is used by other areas of culture. Perhaps even scientists use this ability to a greater extent. After all, literary figures retain many properties of the human originals, while the material point is only a few thought-steps away from nothingness (the empty object). At any rate, the main areas of spiritual culture (science, art, religion) are based on various combinations of simple deformational procedures. These combinations differ but they are also alike – they are conceptually commensurable and ontically homogeneous.

Let us leave the question what the metaphysical significance of these procedures is, i.e. do the "worlds of deformation" really exist, and if so in what sense, or whether we should be instrumentalist about them. What is important here is quite obvious. Science, which idealizes the empirical world, has the same rights as art or religion. The whole spiritual culture grows out of a common core: the ability of our imagination to transcend the empirical. The subareas of culture differ only with respect to how this ability is utilized. Science uses it to reconstruct the empirical world.

Contrary to the current cultural atmosphere in which we are stuck, not for long hopefully, there are no Chinese walls separating science from art or religion. There are no "mountains" on either of the sides either. The past mountains on the side of science, which were raised by scientistic philosophers rather than science itself, are just as illusory as those raised today on the side of art or religion.

4. On the Scientific Community

4.1. *The Problem of the Delay of Social Sciences*

Having waived, or to put it more carefully, having undermined the anti-scientific prejudices a little, we can now return to our question. If the social sciences use the basically same method as the natural sciences, then if the natural sciences could achieve such spectacular cognitive success by using the method, why can social science not do so as well? The surprise that is expressed in this question derives from the assumption that scientific research consists in the transformation of certain raw material into knowledge by means of certain methods. However, such a picture of scientific research is modeled on the picture of a smith who transforms iron into a horseshoe. At the very best, one could use the picture to understand the research of an individual scientist but not science as a whole. Science arises out of the cooperation and competition of individuals. To use Gombrowicz's phrases, it does not take place in them but between them, in a scientific community.

4.2. *Types of Scientific Research*

The assumptions sketched in §2 allow one to distinguish three different types of scientific research. *Applicational research* with respect to a given theory consists in deriving an answer from that theory to a new question, which has not yet been considered. The theory, however, is not questioned at any point. *Correctional research* with respect to a given theory consists in the construction of a new theory that corresponds with the given one. Here the completeness or correctness of non-basic models of the theory is called into question. Correctional work extends beyond the theory but it remains within the bounds of the paradigm initiated by it. *Creative research* in a strict sense extends beyond a given paradigm. It consists in the construction of a fundamentally new theory that dialectically refutes the given one.

Clearly, creative research is more original than correctional research, and that than applicational research. It is also clear that one and the same person can be engaged in research of various types. The research attitude can be determined as the maximum originality that a researcher sets out to achieve. The applicational attitude (a) is exemplified by a scientist who aims only to explain facts without questioning the accepted theory. The correctional attitude (k) is exemplified by a researcher who is prepared to extend the accepted theory in order to offer a better explanation of a given phenomenon. Such a person will propose a new theory that corresponds with the initial theory. Finally, the creative attitude (t) is exemplified by someone who aims to construct a theory that explains facts in a fundamentally different way than the accepted theory.

4.3. *Research Personalities*

The personality of a researcher is determined by the kinds of research attitudes she can adopt in her day-to-day work. If she can adopt only one kind of attitude, she exemplifies the poorest one-component type of personality. The three-component personality is the richest type of personality in our conceptualization.

A complete personality is a three-component personality (a#k#t). It is expressed by the researcher's ability to adopt all three attitudes depending on the needs. *Great creators* not only propose fundamentally new theories but they constantly correct them as well as work on finding new and different applications for them. Newton not only formulated the laws of motion but he was also able to use them to modify Galileo's law of free fall and Kepler's laws of planetary motion. He was also able to apply the law of gravity to explain the phenomenon of tides. Darwin was not only the author of a revolutionary theory but he was constantly correcting it. He was also the author of many applicational studies, among them a rather extensive study on the fertilization of orchids.

This conjunction of creative, applicational and correctional research is understandable and desirable, if one assumes the model of science presented here. In order to (dialectically) reject a theory, one has to know its good and bad sides. The best way to find out what the former are is to apply it to various problems and to correct it effectively. The best way to find out what the latter are is to engage in ultimately failed attempts at correcting the theory. It is thus that real anomalies of a given theory are revealed. They are facts that cannot be explained on the basis of the way of viewing the essence of the phenomenon accepted in the theory. It is only then that the need for a new paradigm becomes something more than a mere desire for originality – it becomes the need of science itself. The person who can answer this need must be capable of carrying out creative research but, at the same time, must have the ability to carry out applicational and correctional research. Thus the most optimal development of a researcher begins with the applicational attitude, through the correctional attitude and ends with the creative attitude, where each of these types of attitudes is permanently internalized and can be used later in case of need.

The remaining types of personality are partial (two- or one-component) personalities. The *creator* of type (a#t) is capable of constructing a new theory and finding applications for it. The creator of type (k#t) is able to construct a new theory and to constantly modify it, but he is unable to deal with everyday applications of it. Finally, the one-component type of creative personality constantly attempts to construct new theories, without caring about their applications or modifications, i.e. about their cognitive infrastructure, so to speak. Such a creator strives for the originality of ideas at all costs, without caring about how they explain facts. History of science provides numerous examples of authors who have made a brilliant discovery, which they soon abandoned because they spent the rest of their lives on fruitless attempts at new discoveries. Some of such authors are not mentioned in the history of science for their discoveries have never made it.

The *applicational personality* is to be found at the opposite end from creative personalities. Someone who is trained in applicational research adopts this attitude permanently and becomes incapable of correcting his theory or of creatively overcoming it. Such a person can only find new applications for the theory. An applicator's personality is in principle a one-component type of personality. This attitude has its place in science and should not be confused with dogmatism. A *dogmatic* is in principle incapable of admitting that the theory she accepts could be false. If a fact that contradicts the theory is found, she will reinterpret it to reconcile it with the theory rather than posing the problem of correcting the theory. The *correctional* personality excludes dogmatism in principle. Correctional research consists in a constant search for facts that contradict a theory and in consequent attempts at modifying that theory. Usually, the corrector's personality has two components: it combines the correctional attitude with an applicational one (a#k). The pure corrector (k) is concerned exclusively with the modification of a given theory. In neither case does the corrector reach beyond the limits of the accepted paradigm.

We can thus summarize the types of research personalities distinguished:

creators:	(a#k#t), (k#t), (a#t), (t);
correctors:	(a#k), (k);
applicators:	(a).

4.4. *On the Necessity of Scientific Schools*

Abstracting from the causes, we can just state a fact: creative personalities in science are rare, and complete creative personalities are even more rare. They are so rare that science cannot afford to wait until the coincidence of luck and proper education yields optimal research personalities. The institution of a *scientific school* is science's way of dealing with the complex tasks of constructing a paradigm, which include not only the construction of a new of theory but also its applications and modifications. The founder of the school is a creative personality but not a complete one. Complete creative personalities usually work alone (Newton, Darwin, Marx). The founder of a scientific school is to some extent incomplete. The school works on the principle that the missing elements in the personality of the founder need to be filled in so that the whole team works like a complete creative personality. Thus the structure of the school: "the founding father," a correctional elite, ordinary ranks of applicators.

Even if the tasks posed by science could be in principle tackled by a single person, schools in science would still be necessary (under the rather obvious assumption that the supply of complete personalities is always smaller than the demand for them). Since it has been the case for a long time now that nobody can construct a paradigm on one's own (one can at most construct a theory that initiates it), the role of the founder of a paradigm comprises necessarily also the role of inspiring and conducting applicational and correctional research. These are only auxiliary reasons why scientific schools are necessary. The fundamental reason is the fact that the collective or social character of doing science lies in the incomplete nature of people doing it.

What is crucial for the development of a given science is the question what types of relationships there are among the researchers and in particular among different schools. Science aims at truth but it can do so only in a collective effort. Let us consider what these relationships must be if science is to progress.

4.5. *Scientific Progress and Its Conditions*

The nature of scientific progress is hotly debated in philosophy of science. For our purposes, it is sufficient to accept the following simple intuitions underlying the concept of scientific progress. Let the basic measure of scientific progress achieved by a given theory be its *explanatory power* (the proportion of facts explained by the theory to the facts in a given domain) and/or the *accuracy* of such explanation (the difference between the theoretical and the empirical value of a given magnitude under explanation and its empirical value). Given a comparable level of accuracy afforded by competing theories, the theory that has greater explanatory power contributes more to overall progress. Given comparable explanatory power of two theories, the theory that is capable of greater accuracy contributes more to overall progress.

Let us note that if t' is a concretization of t, t' approximately explains more facts than those explained approximately by t. The more idealized statement t will approximately explain only classical cases (where the influence of variables omitted in the original idealization is negligible). The concretization approximately explains not only classical cases but also normal ones. It is easily noticed that facts that are jointly approximately explained by t and t' are explained more accurately by t'. In both cases, however, it is taken for granted that the factor that is included in the concretization does in fact affect the investigated magnitude. In such a case – in analogical conditions – the condition of progress within a paradigm is that one theory correspond with another. In the case of an inter-paradigmatic transition the condition of progress would be analogous. A theory that dialectically refutes its predecessor (from a different paradigm) would have to display greater explanatory power. After the development of the paradigm, it would have to explain the same facts at least as accurately as (the best theory proposed within) the rejected paradigm. Scientific revolution that would lead to the loss of the explanatory domain or accuracy would be a dead-end.

In order for those conditions to be fulfilled, scientists and scientific schools must behave in appropriate ways to one another. First and foremost, it is necessary that science be done either by complete creators or by personalities of lower ranks that form complete scientific groups, which consist of the creator as well as correctors and applicators inspired by the creator's way of idealizing phenomena.

The use of idealization presupposes not only imagination (the ability of portraying the world sometimes in entirely different ways from the way we usually picture it) but also respect for this ability whoever possesses it. It is a common interest of the scientific community that there be as many alternative ways of idealizing a given domain as possible. One can never tell in advance which of them will find application and to what problems. And even if it were to turn out that it could not be applied literally, it might still be applied *per analogiam*. Once an analogy is set, it is possible that someone could use it to discover an analogy between analogies. Every discovery of a new internally consistent way of idealizing a given domain persists and profits in further discoveries. Whoever understands this will take delight in discoveries for their own sake – also in those made by others. Thence:

Desideratum of idealization: One should deform empirical reality into theoretical models in as different ways as possible.

However:

Desideratum of concretization: Each of these models must be such that it can be approximated to the empirical world by incorporating the multitude of relevant factors.

The demand of concretization presupposes that one constantly confront the idealized constructs about simple worlds with the empirical facts of our world. It presupposes a critical attitude on the part of the researcher, who has to be to able to reject any view that does not satisfy the criteria of conformity with experience. Again, it is in the interest of the scientific community at large to expose the weaknesses of any views. The desideratum thus presupposes Popper's:

Desideratum of criticism: All scientific theories should be confronted with the totality of known empirical facts; all discrepancies between facts and theories should be exposed.

The exposition of the empirical weaknesses of a scientific theory will be possible provided that theory as well as its critics fulfill certain conditions. An elementary condition that needs to be placed on any scientific theory is Twardowski's:

Desideratum of clarity: Obscure formulations are unacceptable in science. One should strive for the lack of ambiguity and for precision.

This is quite understandable: unclear conceptions could not be subjected to criticism and their explanatory (and predictive) power, not to mention accuracy, could not be determined. To grasp a conception, it is necessary for the critics not only to be competent but also good-willed. Thus the following desideratum, which could be attributed to Ajdukiewicz-Davidson, should complement Twardowski's desideratum:

Desideratum of interpretational charity: If despite genuine efforts to clarify a conception, it remains ambiguous, one should adopt that interpretation under which the conception exhibits the greatest explanatory (predictive) power.

The postulate of criticism presupposes other desiderata, which have been respected in scientific communities since time immemorial and which are so obvious that it is hard to tie them to anybody's name:

Desideratum of egalitarianism in science: Anybody who is sufficiently competent is entitled to engage in scientific criticism and to propose positive solutions of scientific problems without regard to race, political outlook, gender, etc.

Desideratum of freedom of research: Every problem can put forward in science. Every hypothesis or conception can be posed as long as it does not lead to inconsistencies and as long as it is decidable using the methods respected in a given scientific community.

The state of scientific freedom leads to the plurality of scientific schools of thought. Thus Hume's/Feyerabend's:

Desideratum of tolerance: All scientific schools have the right to believe in their paradigm. All competing paradigms have some explanatory power and all have some difficulties (some facts that are incompatible with the theory are known). So it is impossible to claim with any certainty that any single paradigm is true. Since the explanatory position of competing paradigms changes as they are developed, so every school ought to have its chance.

The postulate of developing a given paradigm by means of the method of correspondence requires respect for the scientific tradition, i.e. for earlier attempts to explain the facts. The new theory must first and foremost explain the facts that were explained by a former theory or else it is not even considered as a feasible theory. No theory is so bad that it could ignore facts already explained, and no theory is so good that it would be impossible to explain the facts more accurately. Bohr's desideratum of correspondence is an expression of the respect for tradition and of the requirement that all cognitive potential in a given tradition be utilized:

Desideratum of correspondence: Every theory ought to be perfected by the construction of a new theory that explains what the initial theory has explained and that approximates the idealized constructs of the initial theory even closer to reality.

Some of the rules mentioned above are addressed to different members of the scientific community. The postulate of criticism is addressed to the applicators while the postulate of correspondence to correctors. The following postulate is addressed to creators (in the strict sense):

Desideratum of dialectical refutation: One should always try to deform the phenomena anew by subjecting them to a fundamentally new attempt at idealization. Such an attempt is successful as long as the set of explanations (predictions) is substantially enlarged. It is fruitful if it is possible to propose concretizations that afford the level of accuracy achieved by the best of the alternative paradigms.

The question arises who should realize these desiderata. Clearly, the answer will indicate the adherents of relevant views or their critics. But this answer is not complete. Science is too serious a matter to leave it in the hands of those who believe and who disbelieve. First, it happens not infrequently that there are no rational reasons either in favor or against a view at a given point in time. In such cases, it is necessary to perform theoretical research instrumentally, as it were. In addition, every paradigm leads to a degree of blindness and routine, which are more easily avoided by a neutral person. Frequently, standard research will be much easier to carry out by such a person than by a believer or a critic. Thus, Kuipers':

> *Desideratum of co-thinking:* The adherents of every scientific school ought to help in constructing competing paradigms without regard to their own beliefs. The adherents of different paradigms do not put forward claims on their own behalf but rather put forward conditionals of the form: if the assumptions of a certain paradigm are true then such are their consequences (concretizations, etc.).

Finally, it is a natural human inclination to follow the call of the force. Also scientists are inclined to support those views that have many supporters. Thus Feyerabend's:

> *Desideratum of support for the weakest:* One ought to support the development of the paradigms that have the fewest supporters because the improvement of their explanatory position is particularly difficult and because there is a real threat that they will actually lose any chance at success (assuming that they have such chances).

4.6. *The Ethical Dimension of Scientific Method*

It is not difficult to see the ethical dimension of these desiderata. If in doubt, it is sufficient to confront Ajdukiewicz-Davidson's desideratum with the history of doctrinal religious disputes or with the current ideological disputes, which are full of accusations based precisely on the worst interpretations of the views of the opponents. Or one can confront Feyerabend's desideratum to help the weakest with the frantically followed political rule (not only in totalitarianism) "not to align with the losing fractions." Or one can confront Twardowski's postulate of clarity with the political gibberish we experience all too frequently. And so on. It is sufficient to confront such practices with the practice in the mature sciences to notice that things are quite different at our end. It is not rare in particular in exact sciences to find research conducted on another's behalf, which shows, for example, that somebody else's (sometimes very remote from one's own) theory is consistent. There are also cases where one researcher draws desirable and interesting consequences from another researcher's theory even though the former does not share the latter's assumptions. And so forth. We really have no reasons to be full of complexes. Perhaps it is others who should learn ordinary decency from the conduct of the scientific community.

I do not want to suggest that scientists always behave in this way. Even less do I want to suggest that such a conduct follows from our inborn nobility. Nevertheless, in isolated science (this will turn out to be an important reservation in a moment), conduct that follows such standards is an ideal, to which the day-to-day practice aspires and conforms to quite a significant extent. These standards are respected for the simple reason that deviations from them are penalized by the scientific community. The penalty is the loss of position in the community and, in the extreme case, marginalization or even expulsion. The mechanism is similar to the free market. A salesperson is better off not cheating for her clients will choose competition; as a result, only those who are honest or who learned to be honest remain on the market. Scientific market imposes reliability, critical thinking, clarity, etc. This is why the participants in it willy nilly conform to these demands. The scientific community acts virtuously but not because it is composed of virtuous people. Rather the reverse is true: the community as whole imposes honesty on its members, and those who want to do science must face up to the challenge. This is the case if the scientific community is isolated from the rest of the human world. What happens if it is not?

5. On the Most Important Cause of the Delay of the Social Sciences

5.1. *The Blocking of the Normal Cognitive Development*

If the scientific community is not isolated then the methodological rules that are usually (under the assumption of isolation) respected are sometimes broken under external pressure. Let us look at the phenomenon on the example already mentioned in §2. The condition of scientific progress, as we have seen, is that a (fundamentally) new theory initiate a sequence of variants that correspond with it (a paradigm), and that it be rejected in favor of a subsequent (fundamentally) new theory (new paradigm) in such a way as to preserve the results of the previous paradigm among derivative models (dialectical refutation). In the example in question, one would expect that after a number of improvements (corresponding conceptions), Marx's theory of reproduction will be subjected to a fundamental critique, which will point out that his way of idealizing phenomena of economic reproduction was inadequate. As a result, one would expect that his theory will be replaced by an alternative theory, which will invoke the author's original conception but only in a more distant model. The basic model of this new theory will be based on an entirely new idea of idealizing the process of reproduction. This is what one would expect if only the empirical world, reasoning and nothing else were involved.

Indeed, the Polish economist Rosa Luxemburg has subjected Marx's theory to a severe criticism at the beginning of this century. She has argued that Marx's way of idealizing the mechanisms of economic reproduction was inadequate. Given Marx's assumptions, she maintains, enlarged reproduction is impossible. She reasons thus. Let us assume that there appears a commodity mass on the capitalist market, whose value is $C + V + M$. Whether the owner decides to spend part of the surplus product (accumulation fund M^a) on investments depends on his expectations with regard to whether the commodity mass produced as a result of the investments will be sold on the market. If so, he will invest ($M^a > 0$); if not, he will use the surplus for other purposes, e.g. consumption, and he may even decrease production. As a result, instead of enlarged reproduction, simple or even diminished reproduction may take place. As a matter of empirical fact, however, usually there is such demand. The fundamental question raised by Rosa Luxemburg is this: "Where is this continually increasing demand to come from, which in Marx's diagram forms the basis of reproduction on an ever rising scale?" (Luxemburg 1963, p. 177).[*] In her view, Marx's theory does not provide an answer to this question. She analyzes Marx's reasoning in *Capital* in great detail and shows that it is impossible to find the source of effective demand for investments given Marx's assumptions (that the society under consideration has only two classes and that it is isolated). In general, she shows that Marx did not understand the problem and that he confused it with the problem of the "source of money." She concludes:

> Marx's diagram of enlarged reproduction cannot explain the actual and historical process of accumulation. And why? Because of the very premises of the diagram.[**] (Luxemburg 1963, p. 440)

Enlarged reproduction is thus, contrary to Marx, impossible given his assumptions.

The Polish socialist subsequently constructs a new fundamentally different model of an open capitalist economy, which includes not only

---

[*] *Translator's note.* The English translation is by Agnes Schwarzschild. The translation is available at: http://www.marxists.org/archive/luxemburg/1913/accumulation-capital/ (ch. 7). Access: Aug. 8, 2012.
[**] *Translator's note. Ibid.*, ch. 26.

Marx's two classes but also residues of pre-capitalist modes of production, etc. She shows that this model does answer the question about the source of the demand that enables accumulation. So it also answers the question how enlarged reproduction is possible. In addition, her model explains what Marx's model did though sometimes differently. Whether true or not, this is a beautiful example of theoretical work, which could give rise to one paradigm replacing another paradigm (paving the way to future discoveries by Kalecki-Keynes). The replacement would have taken place if the reasoning about the human world had been conducted at a safe distance from that world . . .

Soon the year 1917 arrived and Marxism became a state ideology of the triple-rule system. All criticisms of Marx, even of his most abstract theory, was out of the question. Only an enemy or renegade, i.e. an even worse enemy, would criticize Marx. "Luxemburgism" was ultimately denounced. As a result, Marx's economic theory was blocked. Instead of undergoing the normal cognitive process, which includes criticism, it underwent canonization – it became impossible even to improve the theory, if it involved showing that the original was in some way inadequate. It was only after the political thaw in 1956 that research on Marx's theory became possible, though only by way of correspondence. Fundamental critique, such as Rosa Luxemburg's, was excluded until the end of real socialism.

In sum, the natural pattern of cognitive development "theory – corresponding variants – new theory – new corresponding variants – . . ." was distorted under ideological pressure. The greatest loss was suffered by Marx's original discovery and his whole economic school, which could not undergo a normal development. This is always the case when a social theory becomes the basis of some social ideology.

5.2. *Blocked Tolerance*

When ideology meddles with science, what is blocked are not only the rather refined mechanisms of scientific progress, such as correspondence or dialectical refutation, but also quite elementary principles. As far as declarations are concerned, everybody will easily agree that in view of the inevitable uncertainty of our knowledge it is impossible to deny any theoretical orientation the status of a contender to truth. In other words, we should grant the right to participate in the scientific market to all serious theoretical options, i.e. those that fulfill all methodological rigors, which are usually accepted (see the desideratum of tolerance above). And so on. How about Marxist economics today? We have seen even in this fragment of the theory of economic growth (incidentally, Marx was a pioneer of the discipline) that it is quite a normal scientific theory. It thus has the same rights to unhampered development as all the other orientations (desideratum of tolerance). In particular, it has the right to expect from other orientations (e.g. the liberal orientation) instrumental cooperation (desideratum of co-thinking). Given how many of its previous adherents have left it, it even has the right to expect special support from the scientific community (desideratum of support for the weakest).

Am I wrong in thinking that my train of thought will be met with some impatience? How can one tolerate, and even support, the ideology of a system that has been crushing all alternative theoretical options, and this by means of secret state police? Well, the aversion to Marxism is quite easily explained and the criticism of the system that it served completely justified. These are ideological matters, however, not scientific ones. Even the most justified of human complaints crumbles when confronted with ideal entities like Marx's abstract economies, which are non-human, neutral to us and far above our sorrows. These entities are what they are: they are either well or badly constructed. Truth lies either with Marx or with Rosa Luxemburg or yet somewhere else. Where it lies, however, does not in any way depend on who, and for what purposes, used the theory. Nor does it depend on who and how greatly suffered as a result. From a purely scientific point of view, Marxism has the same rights as any other theory. Period. The problem is that it is very difficult for us in the humanities to adopt this purely scientific point of view, and so to respect the desideratum of egalitarianism in practice not just in declarations. And then we complain that we are late in our development . . . This is the reason. It is because we cannot rise above our own human skin.

5.3. *Blocked Subject-Matter*

The principle of freedom of research, in particular the researcher's right to undertake any research problem, is even more obvious than the principle of tolerance. It should be emphasized that nobody has the right to impose research subjects on the scientific community or to prohibit the tackling of specific questions. When the scientific community is isolated from the rest of the society, as it is in the natural sciences, freedom of research is by and large in fact respected. However, political sciences, economics, religious studies, sociology, etc. are not so isolated. Thus the normal state of those sciences includes white areas on their subject-matter. Again, real socialism supplies numerous and drastic examples. Limiting oneself to those examples would be, however, tantamount to choosing the easy way in today's Poland. Let us look for more current examples. Soon after the political breakthrough in Poland in 1989, a certain excellent journalist of the old opposition press was nominated as a consul in the USA. His nomination became news, however, because he once happened to write something positive about Scorcese's *The Last Temptation of Christ*. Somebody recalled that and two thousands of "real" Poles signed a protest letter published in an American Polish journal. During a press conference, the question was raised what the stance of the Polish government is on the matter. The speaker of the first government of independent Poland has answered as follows (I cite from memory): "The government is first all surprised. One has to bear in mind that the consul was recommended to this post by his bishop" . . . These are social facts ("pearls" in the jargon), which reveal certain general and fundamental tendencies. It is possible that one of them was revealed to us in that instance. Let us think this through. The minister of a democratically elected government does not say that the consul was nominated according to valid procedures and that everything is fine as long as law is respected. Rather, she defers to an external authority – she leans on the Church. She explains further that this institution holds nothing against the consul. In fact, one may surmise that it even supports him since he was "recommended" by it. The role play by the Church is thus very similar to that played by a certain earlier mono-party. What is the most curious is this territorialism. The consul lives in a certain city, and it is the bishop of that city that recommends him. Is this an accident? Or should we conclude that just as in the real socialist system the recommendations for state posts were divided territorially, so are they nowadays except that a different institution fills the role of the "leading force"?

I do not know. I am just asking. Because I was unable to find an answer, I asked the question of sociologists. In particular, I have asked one of our top experts on the sociopolitical subject-matter, whether anyone followed this "pearl" and conducted some case study. He laughed in response. Nor did he hear of any more general empirical study on the influence of Church hierarchy on the manning of state posts. My conversation with him took place during the fall of 1995. Until then the liberal, especially left-wing, press has multiplied the number of facts, which indicate the political aspirations of the catholic Church hierarchy. There also appeared some theoretical hypotheses, according to which one of the possible developmental paths for our country is catholic totalitarianism. I am not claiming that they are true, they are only hypotheses. We do not know whether they are true or false since empirical research does not provide any grounds for deciding one way or the other. This is the fate of many other problems of this kind. The subject-matter related to the place of the Church in our society is indeed spiked with white areas.

5.4. *On the Dignity of the Humanities*

Well, it is easier to condemn the sociologists or political scientists, but it is more difficult to understand their dramatic situation. Even the very putting forward of a research program, which would comprise empirical studies on the political engagement of the Polish Church, would be met with a wave of political and public outcry from the right to the center (the left would probably exhibit reluctant mortification at the imprudent "waking of the beast"). The ideology of this institution does after all include the thesis that the Church has always been the guard of freedom in Poland. The Pope himself has recently supported the thesis in his Wrocław's speech. Whoever raises the question "Has the Church recommended candidates for state posts in years 1989- . . . ?" admits this possibility as one of the potential answers. In the very act of asking the question, one thus undermines the thesis of the Polish Pope. One, who at this point feels shivers running down the spine as I do, has certainly grasped the cause of the delay of social sciences.

Throughout ages systems, political tycoons, their ideologies have changed. So has the social position of people who reflect on the human world. One thing has been shared by all those people, the humanists: the fear to ask about what really matters. In the humanities, they are dealing with questions, to which politics, ideology, and religion lay claims. They are exposed to threats, but also to temptations, which cannot even be imagined by the lucky ones, who work in different scientific disciplines. From a methodological point of view, their work does not differ too much (give or take a few secondary matters) from that in the natural sciences. Their social position, however, is radically different. They are the addressees of constant claims and postulates, but also of more or less subtle corruption, from the powerful of this world: political powers, political cliques, church hierarchies, mass media. All these forces – the greatest powers of the human world – demand that the humanist support their judgments, that she extols their choice over the rivals, that she crushes their rivals. They expect help in acquiring the reign over souls and bodies. In other words, they want their interests to be warranted with the stamp of truth. As if they did not have enough of their own organizations, disciplined functionaries, well-paid lawyers, devoted journalists, they greedily want science too! It is thus small wonder that the humanities happen to be mired in human interests so deeply that they sometimes lose sight of the right direction. It is small wonder that progress in the humanities is sometimes slower and that its developmental trajectory is more complicated. What is amazing is that, in these conditions, the humanities have managed to tell us so much about the human world. We should respect them for that.

Translated by Katarzyna Paprzycka

REFERENCES

Ajdukiewicz, K. (1934). *Logiczne podstawy nauczania*. Warszawa.

Ajdukiewicz, K. (1960). *Język i poznanie*, vol. I. Warszawa: PWN.

Balzer, W. and B. Hamminga, eds. (1989). *Philosophy of Economics*, Dordrecht: Kluwer.

Brzeziński, J., F. Coniglione, T.A.F. Kuipers and L. Nowak, eds. (1989a). *Idealization I. Poznań Studies in the Philosophy of the Sciences and the Humanities*, vols. 16. Amsterdam/Atlanta: Rodopi.

Brzeziński, J., F. Coniglione, T.A.F. Kuipers and L. Nowak, eds. (1989b). *Idealization II. Poznań Studies in the Philosophy of the Sciences and the Humanities*, vols. 17. Amsterdam/Atlanta: Rodopi.

Brzeziński, J. and L. Nowak, eds. (1992). *Idealization III: Approximation and Truth. Poznań Studies in the Philosophy of the Sciences and the Humanities*, vol. 25. Amsterdam/Atlanta: Rodopi.

Brzeziński, J., B. Krause and I. Maruszewski, eds. (1997). *Idealization VIII: Modeling in Psychology. Poznań Studies in the Philosophy of the Sciences and the Humanities*, vol. 56. Amsterdam/Atlanta: Rodopi.

Egiert, R., A. Klawiter and P. Przybysz, eds. (1996). *Oblicza idealizacji. Poznańskie Studia z Filozofii Humanistyki* 15. Poznań: Wyd. UAM.

Falkiewicz, A. (1989). *Jeden a liczba mnoga*. Wrocław: Wyd. Aspekt.

Feyerabend, P.K. (1979). *Jak być dobrym empirystą*. Trans. K. Zamiara. Warszawa: PWN.

Galilei, G. (1930). *Rozmowy i dowodzenia matematyczne*. Warszawa.

Galilei, G. (1962). *Dialog o dwóch najważniejszych systemach świata: Ptolemeuszowym i Kopernikowym*. Warszawa: PWN.

Garcia de la Sienra, A. (1993). *The Logical Foundations of the Marxian Theory of Value*. Dordrecht/Boston/London: Kluwer.

Hamminga, B. and N. de Marchi, eds. (1994). *Idealization VI: Idealization in Economics. Poznań Studies in the Philosophy of the Sciences and the Humanities*, vol. 38. Amsterdam/Atlanta: Rodopi.

Klawiter, A., ed. (1994). *Understanding Idealization. Theoria* **20**.

Kmita, J. (1998). *Essays in the Theory of Scientific Cognition*. Dordrecht/Boston/London: Kluwer.

Kmita, J. and L. Nowak (1968). *Studia nad teoretycznymi podstawami humanistyki*. Poznań: Wyd. UAM.

Kuhn, T. (1968). *Struktura rewolucji naukowych*. Trans. S. Amsterdamski. Warszawa: PWN.

Kuipers, T.A.F. and A. Mackor, eds. (1995). *Cognitive Patterns in Science and Common-Sense: Groningen Studies in Philosophy of Science, Logic, and Epistemology. Poznań Studies in the Philosophy of the Sciences and the Humanities*, vol. 45. Amsterdam/Atlanta: Rodopi.

Luksemburg, R. (1963). *Akumulacja kapitału*. Warszawa: KiW.

Marx, K. (1955), *Kapitał*. Volume II. Warszawa: KiW.

Marx, K. (1975). Das Kapital. Volume II. Berlin: Dietz.

Nowak, L. (1980). *The Structure of Idealization: Towards a Systematic Interpretation of the Marxian Idea of Science.* Dordrecht/Boston/London: Kluwer.

Nowak, L. (1989-91). *U podstaw teorii socjalizmu.* Vols. 1-3. Poznań: Nakom.

Nowak, L. (1997). *Byt i myśl: U podstaw negatywistycznej metafizyki unitarnej,* vol. 1. Poznań: Zysk.

Popper, K.R. (1977). *Logika odkrycia naukowego.* Trans. U. Niklas. Warszawa: PWN.

Shanks, N. ed. (1998). *Idealization IX: Idealization in Contemporary Physics. Poznań Studies in the Philosophy of the Sciences and the Humanities,* vol. 63. Amsterdam/Atlanta: Rodopi.

Searle, J. (1995). *Umysł, mózg i nauka.* Trans. J. Bobryk. Warszawa: Wyd. Naukowe.

Twardowski, K. (1969). *Wybrane pisma filozoficzne.* Warszawa: PWN.

Wolniewicz, B. (1968). *Rzeczy i fakty: Wstęp do pierwszej filozofii Wittgensteina.* Warszawa: PWN.

Ziembiński, Z. (1967). *Logiczne podstawy prawoznawstwa.* Warszawa: Wyd. Prawnicze.

Wójcicki, R. (1979). *Topics in the Formal Methodology of Science.* Dordrecht/London/Boston: Kluwer.

Leszek Nowak (1943-2009)

THE STRUCTURE OF PROVINCIAL THOUGHT

HALF ESSAY, HALF THESIS (1998)[*]

ABSTRACT. In the essay part, various examples of provincial thinking in Polish culture are recalled. In the thesis part, the phenomenon of provincialism is considered more thoroughly. It is argued that provincialism can be thought of as involving a distortion of a normal division of labor within a scientific school into creators (masters), correctors and applicators. The effect of provincialism occurs when this division is transferred onto whole cognitive communities: some (the centers) play the role of the masters while others (the provinces) are expected to play the role of correctors at best.[**]

1. Gombrowicz, the Provincial "World-Renowned Writer"

Nobody has scrutinized the nature of the phenomenon of provincialism deeper than Witold Gombrowicz. He conveyed to us a lot of painful truths in doing so. It will not be beside the point to remind ourselves of an idea that we have nowadays, in free Polish Republic, repressed out of our consciousness entirely, perhaps precisely because it concerns our current situation. Almost the entire "intellectual class," or at any rate its most visible fragment, feeds itself with the repetition of liberal dogmas, which are rejected only by outcasts, whom the majority, not without a reason, considers to be fanatics. It looks as if there is no alternative but liberalism or nationalist sarmatism. Is this really so? Let us remind ourselves what we have repressed out of our consciousness. Let us remind ourselves of Gombrowicz.

---

[*] The paper appears in English translation for the first time. The Polish original "Struktura myśli prowincjonalnej. Do połowy esej, od połowy rozprawka" appeared in *Przegląd Bydgoski* 9 (1998), 8-20.

[**] The abstract has been added by the editors.

In: Krzysztof Brzechczyn and Katarzyna Paprzycka (eds.), *Thinking about Provincialism in Thinking* (*Poznań Studies in the Philosophy of the Sciences and the Humanities*, vol. 100), pp. 51-66. Amsterdam/New York, NY: Rodopi, 2012.

The liberal Western thought is based on the idea that social life is the result of agreements, contracts, negotiations, undertaken by complete autonomous human individuals. A human is a human thanks to individual conscience – this thought is part of the canon of the today's commonsense rooted way back in Christianity. Witold Gombrowicz has objected to just this canon, which governs European culture:

> Don't we see, at every step, that conscience has almost nothing to say? Does a human kill or torture because he reached the conclusion that he has the right to do so? He kills because others kill. He torments because others torment. The worse act becomes easy, when the path through it has been trodden. For example, in the concentration camps, the path to death was so trodden out that a burgher, who was incapable of killing a fly at home, put people to death with ease . . .
> I kill because you kill. You and he and all of you torment, so I torment . . . sin is inversely proportional to the number of people who engage in it . . . What follows from this is that the spring of action lies not in the conscience of the individual but in the relation that is created between the individual and other people. (Gombrowicz 1973, pp. 60-66)

Yet, says Gombrowicz:

> Our contemporary moral sense has an individualist character. It derives from the sense of an immortal soul, separate for each person. It is thus not suited for the human world, based on so radical a creation of a person by other people. (de Roux 1969, p. 48)

When an individual evaluates, she[*] first and foremost imitates the evaluations of the community. The community judgment is the most powerful factor that affects our conscience. We can disagree with it but only on behalf of another community – either a real or an expected one, which our disagreement will bring to life. Moral sense is built into a person by other people. It is they who form "values" in him, including fleeting values as well as universal ones. Universal values are in particular nothing but constants reflecting the position of an individual in all social configurations. The great moralists – Moses, Christ, the creators of the doctrine of "human rights" – are not legislators who impose their way of looking at the human world onto the peons. Rather, they are great discoverers. They discover and systematize what those peons are inclined to accept and they reject what those peons are inclined to condemn. It is

---

[*] *Translator's note.* I use the feminine and masculine pronouns interchangeably. This is not a form that Leszek Nowak himself often used, though he was alert to sexism in writing and often used the 'he/she' construct.

true that humanity is a matter of "values." What follows from this, however, is only that, and at most that, a human person is called into existence by other people.

This vision of great theoretical worth[1] is still unexplored because our sociologists do not have the courage to put forward their own judgments and prefer to rest content with asking respondents about their opinion on social matters. The theoretical worth of this vision is also exemplified by the fact that it is able to capture quite specific topics non-trivially. An example of such a topic is the problem of provincialism in culture.

This is one of the more important questions Gombrowicz asks: who am I, the Polish writer?

> And what is Poland? It is a country between the East and the West, where Europe begins to fade away, a transitional country, where the East and the West mutually weaken one another. It is thus a country of a weakened form . . . None of the great processes of the European culture has really plowed through Poland: neither the Renaissance nor the religious wars nor the French Revolution nor the Industrial Revolution. Here only quieted echoes could be heard. And the contemporary Russian Revolution has also not been lived through. Poland has only forcibly received its finished effects. Catholicism? The country is admittedly in Rome's orbit, but Polish Catholicism is passive. It consists in the strict adherence to catechism. It has never been a creative cooperation with the Church.
>
> On these plains open to all winds, the great Ridicule and Degradation of the Form had long taken place. It was smeared, it decayed . . . (de Roux 1969, pp. 25-26)

This is not only a description of Poland, it is a description of all provinces – Gombrowicz himself underscores the fact that this is the situation of Rumania, Bulgaria, Argentina, Canada . . . Great literature is not formed on such a ground because even in the hands of the greatest "it was blocked by the tragedy of the loss of independence for one hundred and fifty years" (de Roux 1969, p. 26). The dilemma of the Polish writer was formed against this background. Either he will be true to the Polish problematic but will remain a second-class writer, or he will try to become a world-class writer but carrying with him the baggage of provincial culture he will remain a second-class writer. However hard he tries whether in the direction of the national or the universal, straw will be stuck to his shoes. This is because he, as a Pole, really has straw stuck to

---

[1] For a broader justification of the significance of Gombrowicz's social philosophy, see my (1995).

his shoes. The question is how to remain a Pole and acquire writers' greatness?

There is only one way to do so. Instead of closing oneself off in one's insularity or instead of simulating worldliness, it is necessary to deeply understand and to ruthlessly publicize to the world the truth about oneself, about one's own provincialism.

> If as a human, as a Pole, and as an artist I was condemned to imperfection, there is no point in covering up for it and pretending before oneself and before the world that everything is in a perfect order. To the contrary, it was a matter of honesty, dignity, reason, and, most of all, of vitality to break with the mystification once and for all. (de Roux 1969, pp. 28-29)

The provincial writer has access to a kind of truth that is inaccessible to the lucky inhabitants of the metropolis. She has experienced the clash between the provincial perspective and the universal human problems first-hand, in her own scuffle with the centers of world culture. If she dilutes her experience pretending to be cosmopolitan, she is lost. If she takes advantage of it, she can enrich the culture of the centers with what is invisible from their point of view, with what she herself experiences on her spiritual way to them from the periphery. Then and only then can she succeed.

2. Provincialism and Its Implications

What is a provincial thinker? It is a person who is incapable of venturing his own judgment but who conveys the judgment of the "center" onto an ever deeper province. It is someone who is spiritually subordinated to the "center" but who has pretensions to dominate an even further periphery. Everybody seems to know that Poland is a provincial country and that the Polish nation is a provincial nation, but very few people really understand it. Yet, this is equivalent to the acknowledgement that the Polish mind is held captivated by the Western mind. The provincial thinker is spiritually dependent on the center because he does not dare to accept his own criteria of excellence. Instead, he looks to what is considered to be good There, which he worships, and he disdains what is not recognized There.

Further, the provincial thinker does not dare to compete with the center. He is capable at best of being a representative of the center. He competes with other provincial thinkers but not for excellence, not for a place in the hierarchy of achievements, but for the right to represent the center in the province, i.e. for a place among servants. Finally, if there happens to be, among the provincial thinkers, someone who has the courage to compete for values with the center, she will not only not receive help or even solidarity from the weaker vis à vis the stronger, but she will be actively fought against by her compatriots. The captives by definition never dare to compete with the master. They hate those among themselves who master the courage to start a spiritual rebellion. This is because, by their very existence, the rebels reveal to the captives a deep and unpleasant truth about them.

I live in a typical Polish capital of a voivodeship. The symptoms of provincialism are extremely clear here. Let us limit ourselves to the world of media (I lack the courage to offer a wider description). A well-read newspaper exhibits correspondence from a Warsaw journalist in its daily service. An even better read newspaper builds its position on the commentaries by well-known figures from the capital. A local insert to a central newspaper is very careful not to trespass its role of a local addition to the central spiritual diet even though it has cultivated some really good journalists. (Do they respect the commands of the center? Or perhaps – even worse – do they read off the unspoken wishes of the central and Their Highnesses? The only known weekly, which was created in our city, has been eagerly erasing its roots by relying on the journalists from the capital. Lately, it has even moved to the capital. The direction of promotion in the Polish Third Republic is the same as it was, not to put too fine a point on it, in the Polish People's Republic. Systems change, provincialism is immortal. Why? The cause is our poverty, spiritual poverty – more on the part of the readers than on the part of the editors. The editors of those papers understand very well that even if they find some young local talents, we would not believe in them. How can there be talents on the Warta river? They cannot be true or else they would move to the Vistula river!

The same is true on the Vistula river except that a foreign promotion instead of a capital promotion is at stake. The level of the capital newspapers – to stick with this example – is a little higher than that of local newspapers. It is higher with respect to technique, not with respect to intellectual horizons. Even those papers whose publications could be considered to contain seeds of authentic native thought, such as Rakowski's *Polityka* or later Michnik's *Krytyka*, have gone to the dogs. They feed on current news, mainly related to the "peristaltic movements of the Polish political elites" (to use K.T. Toeplitz's expression), and on the translations and presentations of foreign thought – German in the former case and Anglosaxon in the latter. Among the Poles from the Vistula, only those who have moved to the Rhine, the Seine, the Thames, or the Potomac have lived the high life. The reason why they do not return nowadays is, among others, the fact that they fully understand that, in

Poland, a good Pole is a foreign Pole. A couple of years ago, *Polityka* has run a series of responses by distinguished Polish creators living abroad to the question "Why do I not come back?". Among the curvy idiocies, only Sławomir Mrożek told the truth: I don't come back because you respect me as long as I am here, and I will lose your respect when I come back. Just so. A provincial person downgrades everybody whom she considers to be equal to herself. For deep down she does not respect herself, not even when – and perhaps precisely when – she struts before an external audience. Incidentally, as we know, Mrożek has recently returned. The fact that nobody in the once excellent editorial team of *Polityka* has not had the idea to go to Kraków and ask the Master what has happened that he has come back is a measure of the downfall of the current "illustrated *Polityka.*" Is it an instinctual aversion to the recollection of one of the few significant truths in Polish thought of the 1990s, which would lay bare the cause of the decisive downfall of the only weekly that was once truly outstanding?

I have been speaking about journalism but the point is quite general and, with a few exceptions, applies to other areas of Polish culture. In general, the provincial person feels the greatest aversion to those of his compatriots whose level exceeds his. This is the "Polish hell." The term is deceitful because it has nothing to do with Poles or Poland. It is the hell of all province. The provincial person hates in particular those who, like Gombrowicz, tell her the ruthless truth about herself. So, how could a thought of this caliber be produced in a culture, which has grown on the province of the West, which has made its mark of representing the West in an even deeper province, which feeds itself with its spiritual dependence on the West? This is all the more puzzling since this thought throws that inconvenient truth right into the face of that provincial culture.

I am not trying to be insolent. I am also privy to those heard reflexes. I admit that when I am about to meet an unknown philosopher from Charkov, I am not in the same emotional state as when I am about to meet an unknown philosopher from London. The point is not to condemn but to understand what is the origin of this sticky and ambivalent atmosphere, in which we have been immersed for centuries.

The answer to this question is, I suppose, that we have played the role of a bulwark – first of Christianity, later of Europe. We are proud of our history and of the fact that Poland has guarded Europe from various onslaughts of the always totalitarian East. Our pride is justified, too. It will not hurt us to understand that the costs we have born are not only those that derive from, as we maintain, the "irrational ingratitude" and "short-sightedness" of the West. It is easily noticed that bulwark is a province by definition. The bravest knight from the Kresy, the Eastern borderlands, is mentally a provincial person and is subject to the normal regularities that shape the attitude of an inhabitant of the capital of a voivodeship, for example. He is self-exalting before the "minors," he falls on his knees before the "majors." Pride with respect to some and complexes with respect to others coexist in him and feed one another.

Perhaps even Polish bravery derives to some extent from the expectation of being recognized in the "center," and so from over-eagerness, from the lack of spiritual self-sufficiency? Has this component not been present in the indisputable courage of Chevau-légers, Kresy riflemen, insurgents? Would there have been so much bravery under Somosierra, in the Warsaw uprising, in KOR or "Solidarity," if not for the awareness of the fact that "the (Parisian or New-Yorker) world is looking at us!"? Our bravery is indisputable in its behavioral component. But is it not by any chance a slightly more sophisticated attitude of a provincial person? The inferiority complex of such a person leads him to commit the most courageous deeds, which are undertaken without any calculation or rational reasoning. (KOR, at the time of its historical greatness, is excepted from the latter charge.) This results in the loss of lots of best blood and/or minds – all of this done so as to distinguish oneself and shine before the "majors"? I do not know. I am only asking. I only know that these are important questions and that the impatience with which they are met may be evidence of the truth of a positive answer to them. It may be evidence for the thesis that Polish military courage is largely the result of the lack of civil courage in Poland.

These are not trivial matters. The very same Gombrowicz notices that Polish culture has not yet produced a great philosopher. As a matter of fact, no new philosophy has been put forward in Poland. New variants of already existing philosophies have been proposed (e.g. phenomenology, conventionalism, nominalism). New interpretations of foreign philosophies have been proposed (e.g. anthropocentric Marxism, analytical Marxism, new interpretation of Thomism, the idea of phenomenological Christianity). A new style of doing philosophy has been initiated in Poland (the Lvov-Warsaw school). But nobody among Poles has ever discovered a new hitherto unknown angle, from which one could look at existence and/or the way in which people are anchored in it. At the very best, we have added a couple of less or more brilliant arguments that this cannot be done to the human skeptical treasury. One might add that the arguments are incompatible with the very existence of Meinong, Husserl, Heidegger, Wittgenstein(s), Levinas, etc., to limit oneself to the "philosophical production" of this century only. Of course, we had a couple of great figures but they were from the logical, aesthetic, or historical margins of philosophy, not from core philosophy. The closest to a new core philosophy were the great writers (Leśmian, Gombrowicz), not professional philosophers.

If we take into account the size of our country and its ancient culture, this is quite strange. Great philosophers were born not just by the great and old cultures of Germany, France or England, but also by cultures of countries much smaller (Holland, Denmark, the Czech Republic, Finland) or much younger (the United States) than ours. For sure, there are multiple causes at work. We can suppose, however, that one of the most important among them is the provincial type of mentality specific to Polish culture. We are so small (in our own view and thus in reality) that the "world" must first say that "Gombrowicz (Witkacy, Schulz) is a greater writer" before we dare to notice it ourselves. After the world pronounces this verdict, which is usually late (in fact so late that it is then the "verdict of history" and does not challenge the new world hierarchy, which has arisen in the meantime), we compete with one another in repeating it. In doing so, we are giving evidence of our spiritual poverty for the second time because we are then only reproducing the verdicts of the stronger.

Clearly, this spiritual atmosphere prevents any attempts at a new way of looking at questions as fundamental as the basic problem of philosophy. In this atmosphere, it is wiser (sc. more practical) to exhibit acquaintance of what happens to be fashionable There, and to limit one's own input at most to some commentary of a comparative nature at best (sc. most wisely). And so dozens of papers are published, which present philosophies that are fashionable in the West or, alternatively, which focus on some logical details. I want to be properly understood. Both types of papers are necessary. But neither the former nor the latter (which are of a much greater worth) count as doing philosophy. Preparatory work to (someone's) accomplishment is not after all an attempt at (one's own) accomplishment.

And we need our "accomplishment," some new philosophy, in Central but peripheral Europe much more than do philosophers in the centers of Western Europe. We need it, first and foremost, to overcome the provincialism, in which we are immersed. For our provincialism arises, among others, from the fact that – unlike Wittgenstein, unlike Heidegger, unlike the postmodernists – nobody among Poles has hitherto opened a new trail of thought for exploration. And even if it so happens – as it did several times as far as my memory reaches – that Polish thought produces some single new ideas, which then (apparently independently) also surface in the West with a greater or lesser delay, our compatriots first cannot catch their breath and keep silent for a couple of years on topics quite well known to them from older Polish discussions (the "effect of provincial disbelief"), and then they catch a new breath and throw themselves with homage into Western discussions, citing profusely secondary – in this respect – American or British thought (the "effect of secondary provincialism"). At any rate, Polish thought exploits exclusively foreign trails, which leads to spiritual dependence on the West, and this in turns creates a spiritual atmosphere, which makes the attempts at constructing new philosophy difficult. And so we turn in circles.

As far as Gombrowicz's idea cited at the beginning is concerned, I should add that a project of a new foundation of spiritual culture, which would provide a new look at worlds thoroughly investigated in science and made accessible in art, is needed in Poland particularly nowadays. For it has transpired before our very eyes that when Poles are placed in a really new unprecedented historical situation, they can only respond with the most worn out solutions, which have been tried in the West a hundred years ago. It is quite possible that, from a practical perspective, it was the most beneficial thing to do given all the conditions (including, among others, spiritual provincialism of our country and of neighboring countries). I do not want to engage in a debate at this point, so let us assume that it really was politically the most expedient solution. We should realize, however, that the cultural side-effect of this experimental refraining from thinking in a situation that calls for new thinking is the feeling of idleness and uselessness of truly creative discourse in Poland. Is this not the source of the well-known phenomenon of cultural silence in recent years?[2]

3. On the Structure of Scientific Activity

So much for the essay part. Let's turn to the thesis part. Let us take a look at the phenomenon of provincialism in science. In order to capture it with sufficient precision, we must begin with some general assumptions.

The essence of science is not the generalization of facts but the deformation of empirical phenomena. A theory always rests on idealization. It consists in the introduction of counterfactual *idealizing* assumptions and the construction of the *basic*, the most simplified, *model*. This model comprises *idealizational statements*, whose purpose is to reveal the essence of observational phenomena in a pure form. These

---

[2] The above fragment of the second paragraph is based on an answer provided in the questionnaire conducted by the journal *Edukacja Filozoficzna*, see Nowak (1993).

assumptions are subsequently revoked and the initial model is extended (*concretized*) so as to capture more and more of the phenomena and so as to give a better approximation of registered facts. An *idealizational theory* is a series of such ever more concretized models, beginning with the basic model. Such a theory is accepted if it explains a lot of known facts from a given domain of phenomena to a sufficient degree of approximation. In other words, the theory is accepted if the *approximation* of the theory is accepted.

It is usually the case that a currently accepted idealizational theory cannot explain all known facts. It is then corrected. Upon encountering a fact that is inconsistent with even the most concretized model of the theory, scientists identify the cause of the disparity, i.e. some factor that has not been recognized as yet in the theory. This factor will have to be omitted by means of an additional idealizing assumption. It is only on such an additional idealizing assumption that the claims of the basic model can be accepted. As a result, the basic model becomes even more idealized. The factor will also have to be introduced to the theory. The new idealizing assumption will have to be waived and the claims of the model will have to be concretized by taking into account the impact of the new factor. The result of this process will be a new more adequate version of the initial theory. We will say that the new theory *corresponds dialectically* with the initial theory. What Kuhn calls normal science consists in the construction of theories corresponding with the initial ones. One theory may correspond with the initial theory with respect to one additional idealizing assumption and with another theory with respect to a different idealizing assumption. Since the construction of such theories builds on others (scientists construct new theories that stand in relation of correspondence to theories, which correspond with the initial theory), normal science can be pictured as the construction of a tree, whose root is the initial theory. Such a tree could be called, to use Kuhn's term, a *paradigm*. The relation of dialectical correspondence with a given theory does not extend beyond the paradigm initiated by that theory.

The relation of *dialectical refutation*, on the other hand, does extend beyond the limits of a paradigm. All theories, which comprise a paradigm and which are based on a given idealizational theory, accept its essentialist view of a given domain. In other words, they accept as the principal factors in a given domain of phenomena those factors that were introduced in the basic model of the initial theory. They only supplement the collection of secondary factors. A fundamentally new theory, which dialectically refutes the current paradigm introduces new hitherto unknown principal factors. In this sense, such a theory forsakes the current way of viewing the essence of the investigated phenomena. At the same time, it is a continuation of the old paradigm to some extent. For the theory takes into account some of the factors discovered in the old paradigm. After all, the old paradigm had some explanatory power. When the theory that dialectically refutes the old paradigm is accepted, it once again turns out that it is far from the point of being able to explain all the known facts with sufficient precision and so it must be corrected. In this way, a new paradigm, a new tree of theories that dialectically correspond with that theory, is constructed.

The assumptions sketched above allow one to distinguish three different types of scientific activity. *Applicational research* with respect to a given theory consists in deriving an answer from that theory to a new question, which has not yet been considered. The theory, however, is not questioned at any point. *Correctional research* with respect to a given theory consists in the construction of a new theory that corresponds with the given one. Here the completeness or correctness of non-basic models of the theory is called into question. Correctional work extends beyond the theory but it remains within the bounds of the paradigm initiated by it. *Creative research* in a strict sense extends beyond a given paradigm. It consists in the construction of a fundamentally new theory that dialectically refutes the given one.

Clearly, creative research is more original than correctional research, and that than applicational research. It is also clear that one and the same person can be engaged in research of various types. The research attitude can be determined as the maximum originality that a researcher sets out to achieve. The applicational attitude is exemplified by a scientist who aims only to explain facts without questioning the accepted theory. The correctional attitude is exemplified by a researcher who is prepared to extend the accepted theory in order to offer a better explanation of a given phenomenon. Such a person will propose a new theory that corresponds with the initial theory. Finally, the creative attitude is exemplified by someone who aims to construct a theory that explains facts in a fundamentally different way than the accepted theory.

The personality of a researcher is determined by the kinds of research attitudes she can adopt in her day-to-day work. For simplicity, we can distinguish three kinds of personalities: creative personality, correctional personality, and applicational personality. Usually, scientists belong to one of these personality types. A *complete creative personality* is possible though rare. Such scientists not only propose a fundamentally new theory but spend years working on new and various applications for it as well as on correcting it constantly. A great scientist is able not only to exhibit a creative attitude but he can also adopt applicational and correctional attitudes, if need be. Newton not only formulated the laws of motion but he was also able to use them to modify Galileo's law of free fall and Kepler's laws of planetary motion. He was also able to apply the law of gravity to explain the phenomenon of tides. Usually, however, researchers' personalities are one-dimensional: they either build a new theory, extend an existing one, or apply it without introducing any changes.

Abstracting from the causes, we can just state a fact: creative personalities in science are rare, and complete creative personalities are even more rare. They are so rare that science cannot afford to wait until the coincidence of luck and proper education yields optimal research personalities. The institution of a *scientific school* is science's way of dealing with the complex tasks of constructing a paradigm, which include not only the construction of a new of theory but also its applications and modifications. The founder of the school is a creative personality but not a complete one. Complete creative personalities usually work alone (Newton, Darwin, Marx). The founder of a scientific school is to some extent incomplete. The school works on the principle that the missing elements in the personality of the founder need to be filled in so that the whole team works like a complete creative personality. Thus the structure of the school: "the founding father," a correctional elite, ordinary ranks of applicators. A school is the fundamental cognitive unit, at least in theoretical science.

4. Provincialism in Science

The relationship of domination in science can be expressed thus: x *dominates y cognitively* if y grants x the right to creativity while y grants himself only the right to correct or apply x's ideas (considered to be creative). It follows that someone who is dominated adopts the role of an applicator or corrector of ideas he takes to be dominating. The differentiation of personalities into creative, correctional, and applicational ones takes place spontaneously within a given research community – on the basis of intellectual differentiation. This is quite normal in scientific schools, for example, and in fact it is quite advantageous for science as a whole. Not everybody has a sufficiently independent mind to be able to think of new ideas. On the other hand, everybody should contribute to an improvement of a better justification of new scientific ideas in his own, even if only partial, way. We will call a community with a natural differentiation into creative, correctional, and applicational personalities a "cognitive community." Until this differentiation takes place within a cognitive community, it is quite a natural process. It is like the differentiation of people into those with political aspirations and ordinary (subordinated) citizens, who are quite content that there is someone who takes care of the organization of the social order. Or it is like the differentiation into entrepreneurial personalities and workers, who are content that someone undertakes the risk and puts in the effort to organize production, while they occupy subordinated positions without bearing the burden of responsibility.

The problem begins to surface only when this natural process is distorted by a force factor. One of the most important among such factors is precisely the *effect of provincialism*. It consists in the fact that there appear whole "master" communities, who take themselves to have and who are taken to have leading roles, while the role of all the other communities is downgraded to a supplementary role – of "correctors" community or "applicators" community. In other words, the provincialism effect consists in the replication of the natural structure of a cognitive community at a more global inter-community level, whether in between continents, in between countries, or in between academic centers. This means that the mere fact of belonging to a given research community in a given science is taken to be a sign of the cognitive role an individual is taken to play in that science. If you are from the center, you should create, for you have the right to do so. If you are from a province, you should know better not to be a smart aleck and apply what has been created in the center. If you are from the gray area, you can try to improve (but not change!) "central theories."

The natural division of research personalities in a given cognitive community is thus transferred onto whole communities. A given science is divided into the *central sphere* and subsequent, ever lower, provincial spheres. In contemporary humanities, American universities are the centers, followed by the West-European universities, then Central-European ones, and finally universities located even further to the East than our own. An empirical indicator of this hierarchy is the accepted (and so taken as the universally expected norm) direction of citations. Those from the bottom cite those above, in the last instance – those from the very top. The point is not just the mere direction of citations but the fact that it depicts the hierarchy of world-widely accepted cognitive roles. However, the hierarchy of accepted cognitive roles can (and perhaps even usually does), but certainly need not, replicate the hierarchy of actual cognitive achievements. This is evident if only from the fact that the former is preordained. Your being from Harvard or Sorbonne means, in a given science, that your work contains theories that others, from Ljubljana or Lublin, can only comment: apply to new problems or correct at best, without however questioning their core. On the other hand, nobody from

Harvard or Sorbonne will lower herself to commenting on the work of authors from Ljubljana or Lublin. This is not mere "pathology." It is the norm! It is the more acute the less developed (the less mathematized or, in general, the less technical) the science in question. These artificial hierarchies of influence distort real hierarchies of discoveries in different degrees in different sciences. In mathematics, the degree of distortion is the lowest, while it is the highest in the humanities.

This norm evidently violates the optimal social order in science, viz. the free market. According to this order, every scientist ought to have the same chance of discovering the truth and of being recognized as having discovered the truth if this has indeed happened. Questions of gender, age, academic degree, academic affiliation or national origin ought to be irrelevant. We know perfectly well that this is the most advantageous state for science. After all, the chances of discovering the truth are widely distributed among scientists. In particular, nobody can predict whether truth will not be discovered by someone from Kaliningrad or even from Kazan. Nobody actually knows how many discoveries have been made but not recognized. That this is possible we sometimes discover after long decades. And so only toward the end of this century the economists' world has discovered that one of the most brilliant discoveries in economic theory was made not in the 1940s by Leontieff, the Soviet emigrant in the USA, but at the beginning of the 20th century by someone called Dmitriev. Dmitriev was an economist who wrote only in Russian and whose works were, as it turns out, known to Leontieff. And how many of such accomplished but forgotten (stolen or at least partially recognized) discoveries are still waiting to be rediscovered – nobody knows.

All these issues are painful to the individuals but less so to science as a whole. What is most painful to science is the fact that the artificial hierarchies eliminate, or at any rate substantially decrease, the chances for new discoveries. What is at stake are discoveries not made in the provinces even though they could be made there. For it is understandable that the hierarchy of cultural power depraves provincial scientific communities. Instead of undertaking the task of discovering new theories, provincial scientists are pressured to comment on others' theories. That is what they are awarded for whether by prestige in their own countries or fellowships and invitations from abroad. It is half bad if those who are only capable of filling the role of correctors or applicators do so. The cognitive loss consists primarily in that those very few (there are always very of few of them), who are capable of constructing new theories, usually do not even try to do so for they occupy themselves with the interpretation of "central theories." Though capable of thinking of something new, they work in subsidiary, applicational or at most correctional, roles. After all, the intellect does not always go hand in hand with character or vital energy, which are required in the process of breaking through with one's theory from the bottoms of the hierarchy. Moreover, even if the theoretical discovery does take place, the author frequently has problems with the recruitment of correctors and applicators, i.e. with the founding of a scientific school, which could break through with new ideas on the scientific world market.

Briefly put, science clearly loses out on the division between center and provinces. Why then does the division persist? The answer is obvious and the same as in all kinds of distortions of the order based on the principle of equal opportunities: it is beneficial to those in power, i.e. to those who acquire the chances of becoming even more "equal" due to the division.

Where is the province? The definition has nothing to do with geography: province is where one thinks not on one's own account but on account of another. Provincialism consists in a certain type of academic career (especially in the humanities). It consists in that one does not have the courage to challenge the world but that one makes oneself into a carpet, on which the "world" can march to the Vistula or the Amu-Darya. Provincialism is not defined by one's address but by the type of one's mentality. Who thinks on her own account, who is the center for herself, is not provincial – even if she lives in the countryside in Góra Kalwaria or Vitebsk.

It is not difficult to notice what is the nature of the provincialism of many (though luckily not all) areas of the Polish humanities. It consists in the fact that (sometimes) Polish humanists place themselves in subordinate positions in the world hierarchy. They develop the thoughts of others, not their own. They do not compete with recognized thought options from the centers of their disciplines. They take their own efforts to play the role of commentaries, explanations, or at best completions or applications of others' thoughts. To repeat, the point is not that their work usually has such results since it is natural that, in great many cases, exactly such results will be produced. There is lots of room for various kinds of results in science, including results of lower rank, which can be valuable and sometimes even indispensable. The point is that the very fact that one comes from the province, the very awareness of this fact, leads one to lose the ambition to say anything new. The point is thus not about how many *conceptions* there are in the Polish humanities that could compete with leading conceptions in the world humanities. The point is that there are so few attempts at developing such conceptions. And if they happen to arise, they cannot count on our own endemic intellectual infrastructure. Our own endemic correctional or applicational personalities are busy with developing foreign conceptions. Sometimes even they occupy themselves merely making those ideas accessible on our own market. They popularize science and so do the job of science journalists or research librarians rather than that of scientists. With a few though not very numerous worthy exceptions, instead of fighting our own provincialism, we internalize it.

Conclusion

Is it not a *faux pax* to publish such a text in *Przegląd*? This doubt is natural before reading the text, since the doubt is itself a component of the state of affairs that is our topic. However, someone who had this doubt after reading the text would only prove that he has not understood a single thing. The only rational conclusion that follows from it *de lege ferenda* has Gombrowicz's flavor. *Przegląd* can prove its spiritual independence very easily, viz. by subjecting the very phenomenon of provincialism to scrutiny. The phenomenon is burning in the whole of Poland as well as in the whole of Central and Eastern Europe. Its tantacles are visible in Italy, Sweden, Holland and even Germany . . . Why not accept the challenge? It is a great opportunity precisely for *Przegląd* to initiate serious studies on scientific provincialism. For provincialism is an awful mold, which has grown through the otherwise quite healthy gray cells of this country. And the reaction to these possible future studies – as long they have Gombrowicz's flavor, i.e. are deepened, insightful and surprising – will reveal whose cells this mold has eroded irreversibly.

Translated by Katarzyna Paprzycka

REFERENCES

De Roux, D. (1969). *Rozmowy z Gombrowiczem*. Paris: Instytut Literacki.
Gombrowicz, W. (1973). *Dziennik (1953-1956)*. Paris: Instytut Literacki.
Nowak, L. (1993). Jakiej filozofii Polacy potrzebują. *Edukacja Filozoficzna* **16**, 104-106
Nowak, L. (1995). Gombrowicz i piekło przedmurza. *Bez Dogmatu* **21**, 406.

Leszek Nowak (1943-2009)

MODELS OF SCIENTIFIC RESEARCH (1976)[*]

ABSTRACT. According to the commonsensical model of educating researchers, young researchers must first acquire the knowledge achieved thus far and then solve new problems by developing applications of the accepted theory. This model, which presupposes a positivist theory of science, is incapable of explaining why the major breakthroughs in science have been carried out by young researchers. On the idealizational view of science, it becomes clear that commonsensical model must be rejected and replaced with an alternative, according to which the primary duty of young researchers is to revise the existing theories. It is the young researchers who are usually creative enough, ignorant enough, and exhibit a sufficient degree of nonconformism, to be capable of developing really new scientific theories.[**]

According to a view widely spread in and outside of the scientific community, to do science is to multiply knowledge. In order to do so, one has to carefully study what has been achieved thus far (so as to avoid futile repetitions). Hence the directive for young researchers: study hard so that, having come to know the territories tamed so far, you can spread scientific civilization further. Hence there is a clear distinction between those who are already competent to do science and those who merely aspire to that status. One need not add that masters enjoy a social prestige and privileges (as those "doing" science) while their students carry a marshal's baton in their knapsacks. The division into feudal lords and assistants is only a manifestation of the fundamental differentiation of the scientific community into those who allegedly do science and those who learn how to do it in the future by watching the former.

---

[*] The paper appears in English translation for the first time. The Polish original "Modele pracy badawczej" appeared in L. Leja (ed.), *Rola promotora w kierowaniu kadrą naukową* (Poznań: Wyd. UAM, 1976), pp. 33-41.
[**] The abstract has been added by the editors.

In: Krzysztof Brzechczyn and Katarzyna Paprzycka (eds.), *Thinking about Provincialism in Thinking* (*Poznań Studies in the Philosophy of the Sciences and the Humanities*, vol. 100), pp. 67-74. Amsterdam/New York, NY: Rodopi, 2012.

There is quite a bit of evidence, however, that this differentiation is a myth. There are numerous examples in the history of science, which show that the majority of breakthrough discoveries were carried out by young people, those very same ones who, according to the common view, should only be learning from their older colleagues. If one looks, in turn, at any respected scientific conception, one will realize that it did arise as a result of undertaking new problems that have not yet been solved. Rather it is a result of taking a new incisive look at the simplest problems that had been thought to be long solved. New theories are thus not the products of the missionary drive of their authors to conquer the unknown. Rather they are the results of a revolution that transforms the old theories in such a way as to obtain new solutions of new as well as old problems. Hence every new conception is not an approving extension of old ideas onto new territory. Rather, it is in its inception critical: it is intent on revising the old theory. Small wonder that it meets with criticism from the adherents of the old point of view.

These facts are well known. Perhaps we could say that they are a part of common knowledge about science. Nonetheless they do not lead to a revision of the commonly accepted conception of how science is done. They function more like the anecdotal "exceptions that confirm the rule" than like falsifying facts. The reason for this is, I believe, the fact that they are interpreted in light of the positivist theory of science, which grounds the common view discussed above. If one accepts this theory, these facts are simply unintelligible and as such they do not cause anyone to have any pangs of conscience. During the coffee break, one tells stories of the great scientists making their groundbreaking discoveries in their thirties. After the break, one forces the assistants to do further "studying" and to keep one's own thinking "tied" to what has been read.

The theory of science, which underlies the common conception of the way science is done, is very simple. Its ontological assumptions can be capture thus: the world is a set of phenomena, which consists of constant (repeatable) as well as unique aspects; the constant aspects constitute a hierarchy from the least general to the most general ones. The epistemological assumption, on the other hand, is the following: the aim of science is to find out what the repeatable aspects of phenomena are and to discover the laws of their co-occurrence and succession. The method of discovering what is repeatable in observable phenomena is their cautious generalization, where particular care is taken not to make any mistakes. If the generalizations are sound, they are accepted by competent scientists: they become laws. They are also not revised by later generations of scientists. Rather they are generalized and in turn become special cases of conceptions of a wider scope. As science develops, ever more general aspects of phenomena are discovered.

The common conception of scientific activity is quite natural and understandable on this theory of science. Indeed, if the history of science is the theory of such generalizations of initial theories, then it is quite appropriate to claim that a new would-be scientist must first of all grasp what has been done so far in order to continue, obviously in his mature years, with the dutiful process of generalization. Youth needs to be devoted to education while the creative work (here: generalization) must be left until one acquires an appropriate (here: erudite) training. There are thus two demands unconditionally placed on the young would-be scientist: the demand of erudition and the demand of conservatism, according to which knowledge achieved thus far must not be revised but only generalized. "Study and don't question" are the two bases of the common conception of educating young researchers. Only the latter applies to the educated (the masters).

This conception becomes groundless if one rejects the theory of science in question and replaces it with a different theory of quite contrary philosophical descent. On the ontological assumptions: when a physicist assumes that the gravitational force is more essential to the fall of the bodies than the resistance of the medium, it is not because the gravitational force is a more general aspect of the phenomenon than the resistance exerted by the medium (both these aspects occur always and so are equally general). It is because the gravitational force exerts a greater influence on the phenomenon. The view that is tacitly accepted in science is thus the idea that reality exhibits an essentialist differentiation. For every phenomenon, there are principal (most essential, most relevant) determinants (factors) as well as secondary ones. On the epistemological assumptions: the aim of science is to discover the relations between phenomena and their essences (their principal determinants). In order to realize the goal, the scientist must establish the essentialist structure of a phenomenon (a hierarchy of factors that determine the phenomenon) and then apply the method of idealization. First, she must abstract from all secondary factors in order to capture the relation between the phenomenon and its principal factors in a pure form (the idealizational law). This law is then modified by incorporating the influence of subsequent secondary factors. One thus achieves ever better approximation to observable phenomena. However, the approximation is never completely satisfactory. At first, any deviations are explained in terms of the influence of secondary factors that have not yet been taken into account. One begins to search for such factors. First, new assumptions are formulated, under which the initially formulated law holds. Second, the law is modified so as to incorporate the influence of these factors. The upshot of this process is a new theory, which, as a whole, is incompatible with the old one, but which preserves the main ingredients of the old theory: its laws. If the deviations from observation persist despite many such attempts at developing the old theory, then one slowly comes to believe that the deviations have their source not in the incomplete knowledge about what is accidental (the secondary factors) but rather in a mistaken understanding of the essence of the phenomena (the principal factors). The creator of a new theory proposes quite new factors as the principal ones. In so doing, the old laws are rejected and replaced with new ones, which are subsequently modified to take into account the impact of the secondary factors. And again, deviations from observation, which always surface to some extent and in some degree, are interpreted as indicating the influence of new, yet unknown, secondary factors. The theory is developed so as to take account of such factors but, finally, it is transformed into a theory about new principal factors. (It should be stressed that there is some "overlap" in such transitions for some of the factors of the old theory are preserved in the new theory.)

On such a theory of science, a scientist is obligated to disagree with his predecessors, to revise their theories, to treat those theories as building blocks for the construction of his own distinct theory. On such a theory of science, the model of scientific research is very different from the commonsensical one.

The fundamental virtue of a scientist is, on this model, her creative imagination, which we can identify with the ability to construct new essentialist structures. The greater the difference between the newly proposed and the old essentialist structure, the greater the creative imagination of the scientist. The basic condition required to realize this ability is nonconformism. This is a very important point. When scientists have well-established beliefs about what is essential to the investigated phenomena, when serious, widely respected scientists put forward thesis A about those phenomena, then it is by no means safe to put forward thesis B. The reactions of the group of other scientists can be very diversified but in no way do they resemble the myth of the competent and truth-seeking researchers that the group promulgates with delight. Most frequently, the daredevil is not taken seriously, which can result in civil death within the community, with which his social status, ambitions, and hopes are tied. It is no wonder that it is more convenient and safer to think the way the well-respected do.

At the same time, it is hard to be surprised by the reaction of the group. After all, the people, who comprise it, have frequently devoted their whole lives to meticulous research. They have often relied on assumptions, in which they have such a great faith that they are not even conscious of them. What can their reaction be to someone who claims that the toil of their lives was in vain not because they made some error or because they have deviated from the standards they accept, but because the standards themselves are in principle mistaken – they capture appearances not the essence of things? Those people are simply on the defensive. They defend the positive evaluation of their own life's work and of their social status. Their task is not difficult since it is easy to come up with genuine counterarguments against the new proposal of capturing the essence of the phenomena. After all, that proposal is at first limited to the principal factors. The preparation of an even preliminary list, not to mention a complete one, of secondary factors requires meticulous empirical research. It is thus extremely easy to find substantive arguments against the project of a new theory. Only later can it turn out that facts contradict the theory only apparently, that the deviations from observation can be explained in terms of the influence of secondary factors. At any rate, the fact that the initial project of the new theory shows – because it must show – real gaps and problems only reinforces the beliefs of the defenders of the old order. They take themselves to be justified in speaking on behalf of pure science and standing up against the irresponsible whims of some kids that nobody has ever heard of before.

That is just the point: the kids. After what has been said, it is clear that, statistically speaking, only young people can dare to undertake such a dangerous and risky venture as the construction of a truly new theory, which provides a new way of seeing the essence of phenomena (principal factors). Only they are ready enough to undertake the risk to believe or adopt a point of view only because it is new and fascinating, and contrary to all genuine substantive arguments, which have been amassed by the proponents of the old theory throughout decades. Only they are ready to undertake the toil of empirical research on such fragile foundations. And yet only meticulous empirical research can show that what according to the proponents of the old theory falsifies the new project actually confirms it, if only new secondary factors are discovered that are responsible for the empirical discrepancies due to the consideration of principal factors alone. Only they, finally, have knowledge that is limited enough not to paralyze the freshness of their perspective, not to force old schemata onto facts, and that allows to notice new aspects where the erudites can only see what confirms constructions known from the literature. In other words, statistically speaking, only a young person can exhibit as much intellectual nonconformism and as much vehemence in fighting the old point of view, as is necessary to construct a new theory. And only she has knowledge that is limited enough not to interfere with the new restructuring of factors that are essential for the phenomena under explanation. Thus the fact, which was an idle oddity for the positivist theory of science, becomes fully understandable: the great majority of significant scientific discoveries is made by young people or by people from other disciplines. (The latter share many characteristics with the youth: they are not anchored in the structures of a scientific group, they are social outsiders, and they do not have the disadvantage of having too much knowledge of the discipline.) Some of the sayings that make their away about in scientific circles become understandable as well, e.g.: the old theory dies not because the new one offers conclusive arguments against it but simply because the adherents of the old theory die out (M. Planck). What is difficult to understand, however, is the existing system of educating scientists, which, as we saw, can be captured in the two postulates of erudition and of intellectual conservatism.

This system is often criticized. However, the critiques address the symptoms not the underlying illness: one criticizes the deviations from the system's standards, not the standards themselves. There are lots of complaints that professor X or Y meets their doctoral students once a year or not at all, that one or another adds his name to a paper written exclusively by his assistant, that such and such does not provide her students with access to the laboratory, etc. To assess the commonsensical theory of scientific research, we need to assume that the doctoral advisor is ideal: he honors the copyrights, meets with his doctoral students once a week, helps them in their work, etc. We need to ask whether such an advisor acts in the interests of science, if we accept that science develops according to the principles we have laid out above.

What can the ideal advisor, who respects the assumptions of the commonsensical model of science, offer? He puts forward a new problem and expects of the student that she solve it on the basis of his assumptions. In other words, he expects of her that she derive consequences from his theory (for simplicity, let us suppose that he is the author of the theory that he advocates), which he has not yet noticed. A young researcher is thus someone who extends the current theory. Her research shows that her master's theory has applications reaching further than has been thought so far. It is not hard to notice that, on this "applicatory" model of education, the student works on behalf of the master. She multiplies contributions designed to show that her master's theory is better than he himself thought. I should stress once again that the situation under consideration is optimal on the basis of the commonsensical model. It would be extremely easy to criticize the model by considering situations, when the supervisor does not have any conception and forces his students to engage in quite often quite fruitless collection of facts, when he expects voluminous extracts from the literature (in particular, Western literature), when the doctoral student's only call to glory is that she has mastered the relevant research techniques, but she is unable to use them to propose anything interesting.

Yet, even this optimal model of education is evidently faulty, if the alternative theory of science we have outlined is true. The commonsensical model educates a conservative, someone who does not built new theories but respects the *status quo*. It is a model that enforces acceptance not revision of the current state. It is perfectly suited to the vision of science, according to which science develops by means of the method of small changes with the preservation of the state achieved so far. However, if one rejects this vision and accepts the view that science develops by subsequent corrections of the initial theory, which lead in the end to a new theory with new principal factors, then this model is outright detrimental to scientific progress. The applicatory model of scientific research has a right of place if it is true that the essence of science is stability. The model loses that right if one accepts that the essence of science is a constant revision of the accepted picture of the essentialist structure. The revision consists first in quantitative changes (changes in the secondary factors), which are then transformed into qualitative changes (a revolution, which consists in the introduction of new factors as principal ones, but some degree of continuity is preserved as some of the old factors are still considered to be essential).

Quite a different model of educating researchers is suited to such a view of the development of science. The ideal supervisor imposes only one basic duty on the supervisee: to provide possibly the deepest criticism of his own theory. What is wanted, however, is a scientific criticism not an idle one. In other words, with the discovery of the gap in the master's theory, the young researcher's task only begins. She must find those assumptions, usually unknown to the author of the theory, that are responsible for the problems. In case the problems are less serious, the master's implicit assumptions amount to the omission of some secondary factors. The young researcher then waives those assumptions and modifies the initial theory. She obtains as a result a new variant of the theory, which has the same laws as the initial theory but which is incompatible with the master's theory as a whole. In case the problems are more serious, the implicit assumptions she discovers concern the adoption of an inappropriate set of principal factors. The young researcher then puts forward a new truly original theory, which involves a way of seeing the essence of the phenomena that differs from the one advocated by her supervisor. Since the proposal of a truly new theory is possible only as a result of a deepened understanding of the problems of the old one, the optimal model of educating a researcher should begin with the corrective type of research (say, at the stage of the doctorate) in order to pave the way to the creative type of research (say, at the stage of habilitation).

We should turn to the question how the model of educating researchers we have sketched should be introduced in practice. The worst method one could follow would be to try to convince the masters of its worth in the name of the true interests of science and the like lofty ideals. As is well known, in case of conflict between one's own interests and ideals, it will not be the former but the latter that will find their way to the gutter. I claim, however, that the applicatory model (not to mention its harmful aberrations) will disappear because it must disappear. It stands in the way of the development of human knowledge, which is nowadays becoming indispensable to the survival of our species.

This is most evident in the case of the most developed natural sciences, where the view that scientific research requires creativity not just craftsmanship is paving its way to the mainstream. It is there that an indication of scientific prestige is coming to be the ability to put forward "crazy ideas," ideas that are unbecoming of a "serious scholar," instead of the ability to set commonsensical views into complex formalisms or to use refined techniques to establish common platitudes. It is there that it is most evident that it is necessary to adopt a research system, where the ability to go beyond what is accepted is most highly prized. At the same time, the most developed sciences are the reflection of the future of the less developed sciences. What is necessary for physics today, what is coming to be necessary for biology, will be necessary for the development of the humanities tomorrow. The humanities still largely follow the canons of the positivist methodology. They are still only capable of systematizing the commonsensical view of the world. They are still in the pre-theoretical stage of their development and the only theory, which could transfer it beyond that stage, is for various reasons at a stage, at which its creators left it.

When science will come to face the choice of stagnation or of the abandonment of the applicatory model of educating researchers, the result is easily predictable. For science must develop and so must the models of doing science.

PART II

THE PROVINCE-CENTER DIVIDE
IN CONCEPTUALIZATIONS

Jan Woleński

NATIONS AND PHILOSOPHIES

REFLECTIONS ON TWARDOWSKI'S VIEWS

ABSTRACT. The adjective 'national' in the phrase 'national philosophy' has two meanings. Firstly, disregarding multiethnicity, it can refer to a philosophy cultivated in a country populated by a nation. In such a case, 'national' (e.g. 'Polish philosophy') is nearly synonymous with 'philosophy in a particular country' (e.g. 'philosophy in Poland'). Secondly, 'national philosophy' can refer to a philosophy that displays some characteristic feature of a nation, for instance, its religion or character. This meaning can be illustrated by the phrase 'Jewish philosophy'. This paper focuses on the first meaning in the context of the distinction between philosophical superpowers (national philosophies playing the central role in the development of philosophy) and philosophical provinces (philosophies of a secondary, though sometimes remarkable, importance). Poland is taken as an example. The paper discusses the views of Kazimierz Twardowski, the founding father of the Polish analytic school. He expressed interesting views concerning how philosophy of provinces should be done in order keep its originality.

The adjective 'national' has two meanings in the phrase 'national philosophy'. Firstly, it can refer to the sum of philosophical contributions produced by thinkers belonging to a given nation. National philosophy in this sense is usually related to a country populated by this nation and expressed in a definite native language. Since being a member of a nation and being a citizen of a country are approximately equivalent, we can speak either about X (adjective) philosophy and the philosophy in X (noun); for example, 'Polish philosophy' and 'the philosophy in Poland' are practically co-referential. Some complications appear if a country is multi-ethnic and/or multilingual or if a nation has its own philosophy but it is not organized into a country, but we can skip such cases as not changing very much. The national denomination in the outlined meaning is sometimes replaced by more general labels, like 'African philosophy', 'Arabic philosophy' or 'Islamic philosophy'.

In: Krzysztof Brzechczyn and Katarzyna Paprzycka (eds.), *Thinking about Provincialism in Thinking* (*Poznań Studies in the Philosophy of the Sciences and the Humanities*, vol. 100), pp. 77-86. Amsterdam/New York, NY: Rodopi, 2012.

Secondly, if we speak of Jewish philosophy, it is national in another
sense. Jewish philosophy is intimately related to Jews, their fate and
religion. Jewish philosophy existed long before Israel was established as a
country. It would be a mistake to identify Jewish philosophy and Israeli
philosophy. In Israeli universities, Jewish philosophy is cultivated in the
departments of Jewish Studies, not in the departments of philosophy.
Polish national philosophy is another example of national philosophy in
the second sense. It arose in the 19th century as a philosophical doctrine
expressing the situation of Poles after losing independence of their country
at the end of the 18th century. Polish national philosophy, in particular its
messianic version, focused on a special role of Poland and Polish nation in
the universal history (Poland as the Messiah of nations) and national hopes
and prospects concerning the recovery of independence.

Both of the distinguished kinds of national philosophy should be
understood as ideal types. In fact, actual concrete shapes of philosophies
determined by national factors are fairly complex and vague. In order to
characterize them, we should appeal to various factors, in particular,
general and special philosophical ideas, activities of great thinkers,
cultural traditions, religious factors, geographical coordinates, political
circumstances, etc. (see Collins 2000 for an extensive analysis of
sociological aspects of philosophy). Take religious determinations, for
example. I placed Islamic philosophy among national philosophies in the
first sense, but I pointed to Jewish philosophy as belonging to the second
family. The reason is that 'Islamic philosophy' and 'Arab philosophy'
function as nearly synonymous terms, but we have no counterpart of the
latter label in the case of Jewish philosophy. The cases of Spinoza and Al-
Farabi well demonstrate classificatory problems in this respect. The former
was one of the leading modern rationalists, independently of his relation to
Judaism, but his way of interpreting the Bible belongs to Jewish
philosophy. Although Al-Farabi is commonly considered as the founder of
Islamic philosophy as close to the Muslim orthodoxy (as represented by
Mu'tazilites), he contributed very much as a follower of Greek philosophy
in the Arab world. Thus, rather than a sharp distinction of two kinds of
national philosophy, there is a continuity of positions spanning between
two ideal types as extremes.

Let me consider a concrete example in order to illustrate the previous
general diagnosis. In 1916, a small book on nations and their philosophy
appeared (see Wundt 1916). Wilhelm Wundt, the author of this work,
complained that, due to the war, international scientific co-operation was
practically broken. The book consists of the following chapters: I. The
Rise of the New Worldview; II. Italian Renaissance; III. French
Philosophy; IV. English Philosophy; V. German Idealism; V. The Spirit of

Nations in War and Peace. The first chapter outlines how the new (modern) worldview arose in the Renaissance. Chapter II–IV describe indicated philosophies by their standard historical account, for example, French philosophy is presented by its development from Cartesian rationalism to the materialism of the Enlightenment, but English philosophy is presented as mostly empiricist. The last part programmatically calls to limit national egoism (illustrated by French attitude) and recommends German idealism as a philosophical position important for cultivating real values, especially required for the return to intellectual co-operation. This last appeal is not surprising if we remember about pro-war declarations of German scientists and intellectuals (representatives of other nations behaved in a similar way with a small number of pacifist exceptions, like Russell or Einstein). Clearly, Wundt considers German idealism as *the* philosophy, better than other philosophies for its purely intellectual virtues, but also as something offering the best foundation for social and political interests of mankind.

Wundt finishes the preface in his book with following words (p. 5):

> The author makes clear that in the following presentation he acts as German and that he evaluates philosophical current of other nations from the point of view of German science (*Wissenschaft*). He also stresses that the worldview of German idealism is valid as the philosophy.

One could say that political circumstances caused the fact that Wundt's standpoint contains explicit reference to German idealism as German national philosophy in the first sense. On the other hand, it is quite easy to imagine that Wundt's booklet was published twenty year earlier with minor changes in Preface and the last chapter. Consequently, we would not be surprised that, according to a German leading philosopher, German idealism served as the most valuable philosophy, which provides *the* measure of other national philosophies in the second sense. The word "other" refers in this context to Italian philosophy limited to the Renaissance, French philosophy and English philosophy. We can equally easily imagine similar surveys of philosophy written by philosophers born in Italy, France or England. You can consult multi-volume work of Frederick Copleston, the new edition of Friedrich Überweg's *Grundriss der Geschichte der Philosophie*, *The Cambridge History of Philosophy* or *The Routledge History of Philosophy*. All focus on selected and the same national philosophies. Omitting antiquity, the Middle Ages and the early modern times (up to the end of the 18th century), general textbooks present American, English, German and French philosophy, less frequently Italian philosophy, as constituting the core of the world philosophy. Philosophical ideas coming from other nations are briefly and occasionally mentioned.

Specialized works are devoted to some other philosophical circles, for instance, we encounter books (sometimes fairly extensive) about Russian, Chinese, Arab or Indian philosophy or Jewish philosophy. Books like *Handbook of World Philosophy. Contemporary Developments since 1945* (ed. by J.R. Burr, Aldwych Press, London, 1980) with chapters devoted to many countries are exceptional.

My further remarks are basically limited to national philosophy in the second sense. I have no intention to argue that all nations contributed to philosophy in an equal degree. Clearly, American, English (or rather British in the last centuries), French and German philosophies were (and are) so influential that they deserve to be conceived as of the utmost importance. I also understand that there occur various actual reasons to pay a special attention to Russian philosophy (political importance of Russia), Chinese and Indian philosophy (they are exotic and interesting in comparison with European philosophy), Arab (or Islamic) philosophy (its role as a mediator between Greek-Roman and mediaeval philosophy) or Jewish philosophy (the cultural and religious role of Jews). Sometimes it happens that political circumstances generate a special interest in philosophies existing in some regions that do not necessarily represent superpowers. For instance, the cold war caused a general interest in East-European philosophy as possibly different from that ruling in the Soviet Union. We have also many examples of a special interest in particular achievements, like logical empiricism in Austria, logical ideas created in Poland or analytic philosophy in Scandinavia after 1945. It would not be difficult to demonstrate that almost every country or nation provided an important group or, at least, an individual essentially contributing to world philosophy. However, in general, the philosophical world is divided between superpowers (or centers) and provinces. And this division plays a very important role in production, circulation and consumption of philosophical ideas.

The division between centers and provinces plays not only a descriptive role, but it also has inevitably an explicitly evaluative flavor. The philosophy of the centers is more important in its content as well as in its functions. Philosophical ideas produced by the centers are usually original and determine what is going on in the provinces. Hence, philosophy in the provinces is secondary with respect to the ideas created in superpowers. If something philosophically important appears outside the centers, it is considered to be surprising or unexpected. Reactions to the development of logic in Poland provide a good example. Consider the following statement:

> There is probably no country which has contributed to, relative to the size of its population, so much to mathematical logic as Poland. Leaving the

explanation of this curious fact to sociology of science [. . .]. (Fraenkel, Bar-Hillel, Levy, Van Dalen 1973, p. 200)

This statement is more curious than the fact recorded in it because the size of the Polish population can be hardly regarded as a relevant explanatory factor in explaining why logic developed in Poland so rapidly and strongly. The development of logic in Great Britain, Germany, the United States or Soviet Union is taken for granted, although several factors influenced the process in question. On the other hand, nobody asks why logic did not develop in China or India relative to their enormous populations. Explanations and evaluative qualifications are important because nations are proud of philosophy as a part of their national culture.

The above observations show that the relation between philosophical centers and philosophical provinces is important, although more for the latter than for the former; I will justify this statement in what follows. Hence, the question "How the provinces should consider themselves in relation to the centers?" is legitimate and interesting. This question was considered by Kazimierz Twardowski in Poland. He was a great teacher and organizer of philosophical life in Poland. Twardowski established the Lvov-Warsaw School, the most important group in the entire history of Polish philosophy. In particular, the "curious" development of logic in Poland was a by-product of his activities. It is well documented by Alfred Tarski:

> Almost all researchers, who pursue the philosophy of exact sciences in Poland, are indirectly or directly the disciples of Twardowski, although his own works could hardly be counted within this domain. (Tarski 1992, p. 20)

I would like to present Twardowski's ideas about the place of Polish philosophy in the world-philosophical scenario because I think that they have a general significance, not only for my homeland but for every country (region, nation, circle, etc.) situated outside the centers.

Twardowski insisted that Polish philosophers should know what happens in world philosophy. In order to realize this task, he established a special journal *Ruch Filozoficzny* (Philosophical Movement), which was to provide information about the philosophical life in Poland and abroad. The journal published brief surveys, data and reviews of books and journals, news about conferences and meetings, personal information, abstracts of delivered talks, etc. The first issue appeared in 1911.

The opening paper was written by Henryk Struve (see Struve 1911) and discussed Polish national philosophy. Struve was a very respected Polish philosopher of the turn of the 19th and 20th centuries. He taught in the Main School in Warsaw and the Imperial (Russian) University in Warsaw and published the first specialized academic books about the history of Polish philosophy. Struve began his short paper by indicating that Polish national philosophy should not be understood as national in the first sense (see above). He explicitly excluded Polish national philosophy as something having mystical significance and playing the leading role for the rest of the world. In other words, Struve rejected the understanding of national philosophy, which was popular in Poland in the first half of the 19th century and exemplified by the romantic tradition, in particular, by Polish messianic thought. Thus, Struve himself understood national philosophy in the second sense and compared Polish philosophy with the philosophical thought of other nations, for instance, the English or the French.

In general, Struve had a very high opinion about philosophical skills and potentialities of Poles. On the other hand, he maintained that they are not properly employed. This results in the fact that we have philosophy in Poland as a sum of individual contributions but not Polish philosophy as a synthesis. According to him, philosophy in Poland can be transformed into Polish philosophy (thus he considered the terms 'Polish philosophy' and 'philosophy in Poland' as referring to different things). After this transformation, Polish philosophy could be and should be as important as that developed by other nations, including superpowers (in my terminology). According to Struve, the transformation in question requires that several conditions be met. The internal continuity of the development of Polish philosophy is a necessary condition of its existence. Thus, Polish philosophers should build their views on the work of their predecessors. Struve complained that people working in philosophy in Poland transplant more or less important details coming from the abroad instead of using and developing home achievements. Struve considered philosophy as a web of views. If any personal web is incorporated into historically established system of ideas, the progress of national philosophy becomes possible. Briefly, Struve postulated Polish national philosophy to be a synthesis of the philosophical results created in Poland through the entire history of Polish thought. He was fairly optimistic about successes in this respect and indicated that the first signs of the change required by him "recently" appeared, that is, at the beginning of the 20th century. He finished his paper by expressing the hope that *Ruch Filozoficzny* would help in transforming the philosophy in Poland into Polish philosophy.

Twardowski reacted very soon (Twardowski 1911). His text was not written as a polemic. In fact, he did not criticize Struve but rather considered his own remarks as comments and supplements to Struve (1911). In particular, Twardowski agreed with Struve that Polish national philosophy should not be understood along the lines of Romanticism or something like that. However, after 100 years, we can consider Twardowski's paper as a polemic with Struve's essay; recall that Struve (1911) opened the life of *Ruch Filozoficzny* and was perceived as programmatic. Twardowski came to Lvov in 1895 with a clear intention (see above) to establish a modern philosophical school. He was a very faithful student of Brentano and wanted to base philosophy on the ideas of his teacher. It is very likely that he could have taken Struve's words about transplanting foreign ideas to philosophy in Poland as an allusion to his own program of building Polish philosophy. Moreover, he did not consider the primary task of the new journal *Ruch Filozoficzny* to be a tool for cultivating the continuity of Polish philosophy. Although its role was many-sided, it was certainly one of the basic aims that it open the access of Polish philosophy to the world philosophical novelties. It seems that, in his reaction to Struve, Twardowski wanted to explain his own views concerning the future of Polish philosophy in order to prevent a confusion of his program with recommendations expressed by Struve (1911).

Twardowski did not claim that Polish philosophy should be entirely subordinated to external influences. On the other hand, any home philosophy could not be separated from foreign currents. There is no problem if the issue concerns approximately equal philosophies, like English, French or German (American philosophy was not so strong in 1911 as it is today) because their mutual relations create no danger for their internal development. The situation is different in the case of co-existence of two or more philosophies, of which one is less advanced than another, because the former can be dominated by the latter. Twardowski writes:

> What should be done by a nation lacking its own thought as strongly grounded [as the English, the French or the German nation – J.W.], a nation setting the foundations for home philosophy? Such a nation cannot – even if it would like to do that – surround itself by the Chinese wall in order to block influences from the outside because this would mean to resign from the beneficial effects of co-operation with foreign thought. On the other hand, if this nation opens the door to it, it exposes itself to a danger of muffling its own philosophy. [. . .] It seems that we have only one way out of this dilemma. If, on the one hand, it is impossible to pull out the influences of the more mature philosophy and, on the other hand, every better developed philosophy can dominate our own less mature philosophy, we should remove what is dangerous in the former without losing the profits that it offers. Now, the real danger of influences from foreign philosophy is inherent in its one-sidedness and exclusiveness [. . .] Anytime, if home philosophy is influenced by the only tendency of a distinguished foreign philosophy, English, German or French, this influence must lead to domination. But if foreign influences will act

simultaneously as well as they will cross and link together, this combination results in their mutual balance and weakening of their individual powers. In particular, nothing will be lost from their wealth; on the contrary, they can be complementary. (Twardowski 2011, p. 115)

According to Twardowski, the one-sidedness of foreign influences was responsible for the weakness of Polish philosophy in the past. It is quite possible that he had in his mind Brentano's activity in Vienna. Brentano came to Austria from Germany and wanted to build Austrian philosophy as independent of German thought.

Twardowski returned to these problems in one of his later papers (see Twardowski 1918). In particular, he considered some practical problems connected with academic philosophical education. He argued that Polish students need to be equipped with textbooks of history of philosophy written in Poland because translations from foreign languages do not properly fulfill the required task, viz. the objectivity of presentation. It is typical that German textbooks (the same concerns historical books published by English or French authors) are one-sided because they stress the achievements of their own nations at the expense of other national philosophies (Twardowski could quote Wundt in this respect). Hence, Polish students, if they use textbooks translated from foreign languages, learn the history of philosophy from partial sources because they are colored by a national point of view. Moreover, foreign books usually overlook Polish philosophy. According to Twardowski, a textbook written by a Polish author should and can combine two virtues. Firstly, it should be impartial toward national philosophies and their achievements produced in the centers and, secondly, it should present, although without any exaggeration, the history of Polish philosophy. In general, the proper historical teaching of future philosophers essentially contributes to the situation in which Polish philosophy will not look as a subsequent twin display of English, French or German philosophy.

There were some concrete fruits of Twardowski's program. First of all, he inspired some textbooks of philosophy. *The History of Philosophy* by Władysław Tatarkiewicz, published for the first time in 1931 and subsequently elaborated and edited many times until now, became the most popular Polish textbook of the history of philosophy (even more: the most popular academic textbook in Poland). This work fully realizes Twardowski's postulate that the history of philosophy should ascribe the same measure to particular national philosophies. More importantly, Twardowski recommended to his students that they look for various inspirations and novelties coming from the entire world. Thus, some of his students became interested in logic, others in psychology. And so forth. Twardowski himself was a Brentanist but he did not insist that his students had to follow Brentano in every respect. For example, Kazimierz Ajdukiewicz, one of the most important among Twardowski's students, became interested in French conventionalism in order to weaken the German influences in Poland.

In general, according to Twardowski, any young adept of philosophy should be able to follow all important national currents and to use what is valuable in them. Twardowski hoped that this philosophical practice would initiate a valuable Polish national philosophy and its continuation in the future. The full realization of Twardowski's program was possible in the interwar period, that is, in the years 1918-1939. Although we have no data for a complete evaluation of how far-reaching were Twardowski's ideas concerning the issue under discussion, Polish philosophy had its golden period just between the two world wars.

One can ask whether Twardowski's idea of a balance between the influences coming from various superpowers are still worthy of mentioning. Are they important today? Let me give a concrete example concerning the writing about the history of philosophy. Take a respected *Recent Philosophy, Hegel to the Present* by Etienne Gilson, Thomas Langan and Armand Mauer (New York: Random House, 1962). Its narration is organized nationally. German philosophy occupies (I refer to the Polish translation) 170 pages, French-Italian – 300 pages, English – 150 pages and American – 120 pages. The Vienna Circle is presented in the part devoted to English philosophy. And we find no word about Czech philosophy, Polish philosophy, Russian philosophy or Scandinavian philosophy, although Tarski is mentioned as Polish, Wittgenstein as Austrian, Lenin as Russian or Kierkegaard as Danish. However, Tarski and Wittgenstein are included in English philosophy, but Lenin and Kierkegaard belong to German philosophy, as does Brentano. I do not think that the authors of *Recent Philosophy* do not know that philosophy exists outside Germany, France, Italy, England or the USA, but this fact has no particular significance for them. In other words, philosophy is entirely exhausted by the ideas produced in superpowers.

One could maintain that globalization makes Twardowski's worries and recommendations less important. International co-operation is easy, the same concerns access to books and journals, internet provides quick and effective tools of communication, and we have English as a new *lingua franca*. However, I do not think that the situation changed radically since 1911. The very division between philosophical superpowers and philosophical provinces still exists and perhaps is even deeper than a hundred years ago. The textbook written by Gilson, Langan and Maurer is not secluded. To avoid possible misunderstanding, I do not claim that Polish philosophy should occupy as much space as American, English,

French or German philosophy but I only propose that it be mentioned as an example of something valuable, if there are reasons for that, even if it is provincial.

Since these remarks concern many countries, they are addressed more to philosophers working in the provinces than to the representatives of the superpowers. Although the latter are more or less sympathetic to our interests, but their attitude is (statistically speaking) decisively paternalistic. I asked once why nobody from Central and Eastern Europe (excluding Germany) was appointed as an invited speaker or a member of the program committee at one of the most important philosophical congresses. The answer formulated by the president of the association organizing this congress (of course, he was recruited from a superpower), was that such appointments were motivated by political reasons in the past but they lost their significance after 1989. No comments.

University of Information, Technology and Management
ul Sucharskiego 2,
35-325 Rzeszów
Poland
e-mail: wolenski@if.uj.edu.pl

REFERENCES

Collins, R. (2000). *The Sociology of Philosophies. A Global Theory of Intellectual Change*. Cambridge, MA: The Belknap Press of Harvard University Press.

Fraenkel, A., Y. Bar-Hillel, A. Levy, D. Van Dalen (2003). *The Foundations of Set Theory*. Amsterdam: North-Holland Publishing Company.

Struve, H. (1911). Słówko o polskiej filozofii narodowej [A Word about Polish National Philosophy]. *Ruch Filozoficzny* **1**, 1-3.

Tarski, A. (1992). A Letter to Otto Neurath, 25 IV 1930. German original and English translation by Jan Tarski. *Grazer Philosophische Studien* **43**, 10-12, 20-22.

Twardowski, K. (1911). Jeszcze słówko o polskiej filozofii narodowej [Another Word about Polish National Philosophy]. *Ruch Filozoficzny* **1**, 113-115.

Twardowski, K. (1918). O potrzebach filozofii polskiej [On the Needs of Polish Philosophy]. *Nauka Polska* **1**, 129-163.

Wundt, W. (1916). *Die Nationen und Ihre Philosophie*. Leipzig: Alfred Kröner Verlag.

Giacomo Borbone

FROM COSMOPOLITISM TO NATIONAL-POPULAR CULTURE

GRAMSCIAN ATTEMPT AT OVERCOMING PROVINCIALISM

ABSTRACT. Circulation of ideas among philosophers is the core of Philosophy itself. The lack of this circulation can lead to obscurantism and cultural provincialism. The latter, for instance, afflicted Italy during the first half of the 20th century because of the close-minded neo-idealism of Croce and the mutual indifference of science and philosophy. Antonio Gramsci tried to overcome the problem of provincialism. In this essay, I explain how he attempted to overcome it. I focus on his conceptual categories like hegemony, organic intellectual, national-popular and so on.

1. Introduction

The problem of cultural provincialism afflicted Italy during the first half of the 20th century for several reasons among which there are: the close-minded neo-idealism of Benedetto Croce, the mutual indifference of science and philosophy and the political power of fascism.

All these problems were well understood by the Italian philosopher Antonio Gramsci who tried to overcome the problem of provincialism through a deep historical and philosophical-political study of Italian culture. According to Gramsci, the roots of the coeval Italian situation could be found in the Cosmopolitism that followed the fall of the ancient Roman Empire and that extended itself to Fascist Italy.

This Cosmopolitism did not allow the Italian political class and culture to give birth to a national-popular culture just because Cosmopolitism was far away from people, from masses. The consequence was the provincialism of Italian culture and its most important representative was the Italian philosopher Benedetto Croce. Gramsci, as is well-known, criticized the intellectual role of Croce in his *Prison Notebooks*, postulating the necessary passage from a cosmopolite culture to a national-

In: Krzysztof Brzechczyn and Katarzyna Paprzycka (eds.), *Thinking about Provincialism in Thinking (Poznań Studies in the Philosophy of the Sciences and the Humanities*, vol. 100), pp. 87-102. Amsterdam/New York, NY: Rodopi, 2012.

popular one, closer to concrete historical events, to reality and to people concretely acting.

In this essay, I will try to explain how Antonio Gramsci attempted to overcome provincialism through his conceptual categories like the ones of *hegemony, organic intellectual, national-popular* and so on.

2. The Relevance of the *Cultural Factor*

Gramscian theses on the role of the political party conceived as a *collective intellectual* and *Modern Prince* and on the role of the so-called *organic intellectuals,* who must not remain linked to the mere theoretical level but plunge in the practical life acting as cement for the constitution of a new *historical bloc,* are well known.

Gramscian work is sparkling with political contents just because there is no doubt that the main theme of his reflection, specially the one present in his *Prison Notebooks,* is essentially of political nature, dealing with the problem of the conquest of hegemony (conceived first of all in terms of a cultural direction) by the working class.

Gramscian reflections contain a lot of original elements deriving first of all from a strong anti-dogmatism that finds its roots in Antonio Labriola's philosophy of praxis. In fact, in the area of Antonio Labriola's reflections and political action, the cultural factor, conceived not in an abstract way but always as a vehicle of social progress, plays an important and essential role. Culture, considered as an emancipator factor, was able to give to the working class the chance to achieve the critical consciousness that could put it into the concrete possibility to fight for its own freedom and rights that, according to Labriola, were alienated by the capitalistic system. The path was not simple. In fact, Labriola found himself in a condition of isolation.

Labriola's intuition found a validation in Gramscian *Prison Notebooks,* where Gramsci used the Leninian concept of hegemony though he re-elaborated it. According to Gramsci, hegemony is a pre-condition for the catch of power, linked to an *intellectual and moral reform* postulated by Gramsci himself. But the originality given by the *Prison Notebooks* consists of conceiving the *consensus* just *before* the catch of power through the gradual conquest of the so-called *earthworks.* In this respect, these Gramscian reflections are very emblematic:

> The methodological criterion on which our own study must be based is the following: that the supremacy of a social group manifests itself in two ways, as "domination" and as "intellectual and moral leadership." A social group dominates antagonistic groups, which it tends to "liquidate," or to

subjugate perhaps even by armed force; it leads kindred and allied groups. A social group can, and indeed must, already exercise "leadership" before winning governmental power (this indeed is one of the principal conditions for the winning of such power); it subsequently becomes dominant when it exercises power but even if it holds it firmly in its grasp, it must continue to "lead" as well. (Gramsci 2001, pp. 2010-2011)

As we can notice, Gramsci used to give a lot of importance to the super-structural element, both in the political and philosophical reflection, giving proof of a strong anti-dogmatism that can be ascribed to Labriola.

But these theoretical Gramscian positions must not be conceived as a Gramscian removal from Marxism[1]; just the opposite, Gramsci was and remained always a Marxist but a Marxist who, following Labriola's great lesson, used not to interpret Marx and Engels's thought on the base of a rigid and deterministic structure/super-structure relationship (about the Gramscian anti-dogmatism, see Gramsci 1917). As Rupert said, "Gramsci accepted in broad outline Marx's analysis of the structure and dynamics of capitalism but was unwilling to embrace the more mechanical and economistic interpretations of Marx circulating in the international socialist movement" (Rupert 2005, p. 487).

3. Methodology or *Weltanschauung*?

In the *Prison Notebooks*, the most quoted intellectual is the philosopher Benedetto Croce but the real source of Gramscian anti-dogmatism and philosophy of praxis is Antonio Labriola, who exerted a deep influence on Gramscian conception of Marxism. For instance, Gramsci stated that

> In reality Labriola, who affirms that the philosophy of praxis is independent of any other philosophical current, is self-sufficient and is the only man who has attempted to build up the philosophy of praxis scientifically. (Gramsci 2001, p. 309)

---

[1] Norberto Bobbio put in evidence the difference between Marx and Gramsci about the way of conceiving the concept of civil society: for the first, it is an element of the *structure* (the reign of economic relationships); for the latter, an element of the *super-structure*, (the reign of cultural-ideological relationships). Bobbio, after having demonstrated the priority given by Gramsci to the super-structural moment, writes that the re-evaluation of civil society is not what links Gramsci to Marx but, at the opposite, is what distinguishes the first from the latter. This divergence, according to Bobbio, does not put Gramsci out of Marxism because all Marxist theories accept the fundamental dichotomy between structure and super-structure, and Gramsci accepts this dichotomy. See Bobbio (1977).

This quotation is very relevant because of a novelty element scarcely considered in the Marxist literature, i.e. the philosophical independence of historical materialism.[2] This reflection takes away Marxism from the ones who used to link it, with a twisting interpretation, to Hegelianism or to Immanuel Kant's philosophy of morals. There is no doubt that both Hegel and Kant played an important role in Marxian philosophy but the innovative concepts like the ones of historical materialism, structure/super-structure relationship and the relevance of the economic factor, cannot be ascribed to the influence of Kant or Hegel.

Anyway, Labriola was interested in the effectuality of history but a history considered as free from the metaphysical fetters of the dualistic structuring of the problem in terms of ideal history and real history. Contrary to this way of facing the problem, Labriola (1976, p. 475) wrote that history is based on the process of formation and transformation of society (as Vico would say, history is a human product).

Effectuality of history and history conceived as a human product are the starting points for the analysis of historical processes but right here we can find an essential difference between Labriola and Gramsci. In fact, as we have seen, both Labriola's anti-dogmatism and anti-economism, were accepted by Gramsci but while, according to Labriola, Marxism had to be *naturalized*, in Gramsci there was not a similar cultural operation. What does 'naturalized' mean in Labriola's reflection? It means that Labriola used to conceive Marxism as a *method* while Gramsci intended it as a *Weltanschauung*.

Labriola realized that Marx used a specific method in his analysis of *capitalistic modes of production*, i.e. the method that the Pole philosopher of science Leszek Nowak and his Poznań School of Methodology called *idealization* and *concretization*.

The main role of this scientific approach, which Leszek Nowak reconstructs starting from a methodological study of Karl Marx' and Galileo's works and that he re-elaborates in a creative way and with a close comparison with the most significant conceptions of contemporary epistemology (Carnap, Hempel, Popper, etc.), takes its conceptual nucleus from the Marxian need (in turn taken from Hegel) to separate essence from appearance, to catch what is more essential in a phenomena, which is, according to Nowak, the fundamental aim of science.

---

[2] On the theoretical autonomy of Marxism, Giuseppe Cacciatore writes: "The claim of "theoretical" autonomy of Marxism, in Labriola assumes the meaning of a defense of its scientific status" (Cacciatore 2005, p. 23).

But this need can be completely developed only if we turn our gaze toward what constitutes Nowak's theoretical starting point, both simple and innovative, viz. his thesis about the difference between abstraction and idealization. In contrast to what has been usually been maintained by inductive philosophies or even by the positivist and post-positivist ones, it is idealization not abstraction, according to Nowak, that is the core of scientific method. The difference between the procedures is explained by Nowak in the following way:

> A scientific law is basically a deformation of phenomena. It resembles much more the logical structure of a caricature than that of the generalization of facts. The crucial point for a proper understanding of the Idealizational procedure is that it differs fundamentally from that of abstraction [. . .]. Abstraction, i.e. the omitting of properties, leads from individuals to sets of individuals (and from sets of individuals to families of sets, etc.). Idealization does not do this. Omission of the dimensions of physical bodies does not yield any set of physical bodies but the mass-point. Abstraction is generalization. Idealization is not. (Nowak 1992, pp. 10-11)

The difference between abstraction and idealization consists in the fact that, while the first is applied by human intellect obtaining universal concepts from the knowledge of particular objects (by the generalization of empirical facts), by "idealization," we proceed "enclosing between parenthesis" some aspects of phenomenal reality that we consider secondary, in order to take into consideration the essential factors of the phenomena under investigation (see Borbone 2011a, pp. 227-252; Borbone 2011b, pp. 34-48; Coniglione 2010). Antonio Labriola noticed the idealizational nature of Marxian economic theories, in fact he conceived Marxian *Capital* as an idealized model. However, Gramsci abandoned the path followed by Labriola and instead preferred to conceive Marxism only as a *Weltanschauung*. This is, maybe, the weaker aspect of Gramscian philosophy, even though he understood the idealizational nature of scientific theories (see Coniglione 1986; 2008, pp. 87-88). Gramscian philosophy was strongly linked to the political problems that afflicted coeval Italy. One of the main problems that according the Italian philosopher had to be overcome was provincialism. That is why, Gramsci preferred to develop a *Weltanschauung* in order to analyze political, moral and economic Italian situation.

But the problem of provincialism cannot be fully understood without the study of the main Gramscian philosophical themes and categories.

4. Hegemony, Organic Intellectual and Historical Bloc

Before providing some solutions to the most pressing problems of Italian
political and economical situation, Gramsci realized that he needed a
strong categorial apparatus, essential in order to understand and, then,
overcome problems like provincialism, the isolation of the working class,
the overcoming of the traditional intellectual, and so on.

The conceptual frame of Gramscian reflection is the so-called
philosophy of praxis, which is, according to the Italian philosopher, an
absolute humanism of history, an absolute historicism "which represents
the climax of modern history" (Pagano 2007, p. 4). Gramsci, in this
respect, comes back to the Marxian *Thesen über Feuerbach,* where the
German philosopher wrote that

> The chief defect of all hitherto existing materialism – that of Feuerbach
> included – is that the thing, reality, sensuousness, is conceived only in the
> form of the *object or of contemplation* but not as *sensuous human activity,
> practice,* not subjectively. Hence, in contradistinction to materialism, the
> *active* side was developed abstractly by idealism – which, of course, does
> not know real, sensuous activity as such. Feuerbach wants sensuous
> objects, really distinct from the thought objects but he does not conceive
> human activity itself as *objective* activity. Hence, in *The Essence of
> Christianity*, he regards the theoretical attitude as the only genuinely
> human attitude, while practice is conceived and fixed only in its dirty-
> judaical manifestation. Hence he does not grasp the significance of
> "revolutionary," of "practical-critical," activity. (Marx [1845] 1997, p. 82)

Gramsci deeply metabolizes this Marxian conception of praxis to the
extent that the content of this Marxian quotation will have, for Gramsci, a
decisive importance on his way to conceive not only the relationship
materialism/idealism but also history. Gramsci understood very well that
as idealism hypostatized *spirit*, old materialism hypostatized *matter*:

> in the philosophy of praxis quality is also connected to quantity and this
> connection is perhaps its most fertile contribution. Idealism, on the other
> hand, hypostatizes this mysterious something else known as quality, it
> makes it into an entity of its own, "spirit," just as religion had done with
> the idea of divinity. But if the notion of quality is a hypostasis in religious
> thought and in idealism, that is to say an arbitrary abstraction rather than a
> process of analytical distinction necessary for explanatory purposes, then
> the same is true in the case of vulgar materialism, which "divinizes" a
> hypostasis of matter. (Gramsci 2001, p. 1447)

Gramsci thus aimed to overcome the classic epistemological position
merely postulating the existence of a subject and of an extern and
independent reality. The human subject is not passive in front of the object

but active through a modification of his surrounding environment. In order to understand this point better, it can be useful to explain the difference between *Objekt* and *Gegenstand*. Marx conceived the first as an object that stands in front of the subject passively, where the subject has no relationship with it; the latter, instead, is conceived by Marx as what the subject creates and puts in front of himself in an active way; furthermore, this aspect of Marxian reflection finds its roots in Johann Gottlieb Fichte's *Wissenschaftslehre* characterized by the dynamic Ego/non-Ego (see Fichte [1794] 1845; about the idealistic roots of Marxian thought, see Borbone 2012).

Gramsci inherits this conception of praxis canalizing it in the political sphere because, according to him, theory (conceived as a *Weltanschauung*) must be translated into a conscious political action. Thus, the identity between philosophy and politics reaches its climax.

Another identity established by Gramsci is the one of philosophy and history, to the extent that he talks, in this respect, of a kind of *bloc*:

> The philosophy of an age is not the philosophy of this or that philosopher, of this or that group of intellectuals, of this or that broad section of the popular masses. It is a process of combination of all these elements, which culminates in an overall trend, in which the culmination becomes a norm of collective action and becomes concrete and complete (integral) "history." The philosophy of an historical epoch is, therefore, nothing other than the "history" of that epoch itself, nothing other than the mass of variations that the leading group has succeeded in imposing on preceding reality. History and philosophy are in this sense indivisible: they form a bloc. (Gramsci 2001, p. 1255)

And it is right history that acts as a test-bed for Gramscian theories of *hegemony* and *historical bloc* just because Gramsci realizes that the backwardness of coeval Italy has remote roots, i.e. the crisis or dissolution of the Roman Empire. According to the Gramscian historiographical reconstruction, from the crisis of the Roman Empire, through the eighteenth century, Italian intellectuals adopted a cosmopolite character that prevented them from exerting a *national-popular function*. The *national-popular* category is used by Gramsci mainly for the analysis of literature (for instance, some very famous pages from the *Prison Notebook* are devoted to Luigi Pirandello or to Alessandro Manzoni). However, Gramsci extends this category also to history and politics. Very emblematic, in this respect, is the case of Italian Risorgimento. In fact, while the most part of Italian historiography used to talk of the Italian Risorgimento as a *national culture*, Gramsci, by contrast, stated that it was not a national culture but a *pure verbal illusion*. In fact, Gramsci asks himself:

Where was the basis for this Italian culture? It was not in Italy; this "Italian" culture is the continuation of the mediaeval cosmopolitanism linked to the tradition of the Empire and the Church. Universal concepts with "geographical" seats in Italy. The Italian intellectuals were functionally a cosmopolitan cultural concentration; they absorbed and developed theoretically the reflections of the most solid and indigenous contemporary Italian life. (pp. 1361-1362)

According to Gramsci, the absence of the national-popular element in the Italian cultural background prevented the formation of a leading class linked with the masses and, as a consequence, the formation of the organic intellectual. This situation generated an anomaly in the Italian State, i.e. the presence of a particularistic and provincial mentality still deeply rooted in the Italian habits, in the way of thinking and in the way of acting.

Therefore, Italian Risorgimento, which Gramsci interpreted as a *missed revolution*, facilitated the birth of capitalism and of bourgeoisie. At the same time, the Italian capitalism

conquered the power by following this line of development: it subjugated countries to industrialized cities and it subjugated both central Italy and southern Italy to the North. (Gramsci 1966, p. 40)

While Benedetto Croce considered Italian Risorgimento as the "masterpiece of the liberal-national movements of nineteenth century" (Croce 1993, p. 275), Gramsci, instead, stated that in reality Italian bourgeoisie only unified Italian people *territorially* but not *spiritually* and *economically*.

This goal, left undone by Italian Risorgimento, in Gramscian conception could be fulfilled only by the working class:

For sure, only the working class, by snatching from capitalists' and bankers' hands political and economic power, can be able to solve the central problem of Italian national life, the Southern Question; for sure, only the working class can bring to an end the laborious effort of unification that begun with Risorgimento. (Gramsci 1966, p. 40; see also Gramsci 1920)

In his reflections on Italian Risorgimento, Gramsci

observed how the democratic republican leaders around Mazzini and the Action Party failed to generalize their struggle beyond the radical bourgeoisie and win the support of the peasantry. They were thus subsumed and defeated by the Moderates under Cavour, who were able to construct a hegemonic alliance of the bourgeoisie with the southern landowners, an alliance whose continuation was secured in the state through transformism (*ad hoc* ministerial coalitions) and by the economic subjection of the South to the North in a colonial relationship offset for the big landowners by protectionist policies. This "chapter of past history" bears on the concrete relations of force in the 1920s and 1930s because of its historical parallel with the PCI at the time of the rise to power of fascism. This party too had failed to become hegemonic because of its inability to carry out the Jacobin task of linking countryside to city – peasantry to proletariat, south to north – to form a national-popular collective will. In its place, the Fascist Party had carried out a reform revolution or passive revolution, based on the defensive, transformist alliance of the industrial bourgeoisie, big landowners and petty bourgeoisie, which involved no fundamental reorganization of the economic structure – only its technical modernization along rational "Fordist" lines – coupled with an increase in state coercion and the securing of mass popular consent in civil society. (Forgacs 1999, p. 213)

As we have seen, the overcoming of Cosmopolitism through the *national-popular* category was useful for Gramsci in order to interpret some crucial moments of Italian history: in particular, the Italian Risorgimento conceived as a *passive* or *missed revolution*.

But what clearly emerges until this point, is that the Cosmopolitism criticized by Gramsci is the one of the Italian traditional intellectuals, i.e. the one linked to Church or to Roman Empire: a cosmopolitism that, in Italian History, has always been an insubstantial myth, without any link with the present reality or in evolution. Moreover, this cosmopolitism fueled the traditional Italian people's subversivism.

This specification is necessary in order to clarify that Gramsci was not contrary to a modern form of cosmopolitism. Quite the opposite, he aimed for a kind of cosmopolitism deeply rooted in the actual characteristics of Italian society. Which ones?

The first one was the necessity of overcoming the close-minded politics, the autarchy that was affirming itself during the political power of Fascism. Gramsci realized that while economic life has got, as its premises, internationalism or cosmopolitism, State life, instead, developed itself into a form of nationalism. This particular form of nationalism, according to Gramsci, is the source of the actual crisis because it does not comprehend that the world is a unity (in modern terms, a globalized entity).

But there is a more general reason for that. The Italian people, among others, is more interested in obtaining a modern form of cosmopolitism just because the Italian labor-man, with his work, fecundated all the world. He is thus the natural bearer of cosmopolitism able to reconstruct the world unitarily (and not to dominate it in a hegemonic way embezzling the fruits of one's labor). So, we are talking about the working people's cosmopolitism and of its intellectuals (organic, new, and not traditional), a cosmopolitism that, according to Gramsci, must become an operating internationalism.

There is also a tendency to a cosmopolitism of capital but it is clear that this kind of cosmopolitism just because it aims to extend the exploitation of man by man on a planetary scale, cannot be the proletarian one. Even though, it is not out of question that Americanism and Fordism could be utilized by the proletarian State in order to create that new-man useful for the planned production.

According to Gramsci, the formation of an Italian national consciousness must occur through the overcoming of both municipal particularism and catholic cosmopolitism, at that time strongly linked. This overcoming meant the creation of a secular man in combat with Catholicism and not only set apart from it.

Obviously, a modern cosmopolitism does not prevent the formation of an Italian national consciousness, and in this sense Gramsci cites the example of Italian Jews. Jews' religious cosmopolitism becomes a particularism within the National States but they overcame it and tried to build an Italian national conscience between the 17th and the 19th century (as the Piedmontese and Neapolitans were doing at the same time).

Gramsci, in this particular and original re-reading of Italian history, includes his conceptual apparatus or categorial luggage, that is also the theoretical skeleton of his philosophy of praxis, and it is right in the light of this conceptual and categorial frame that Gramsci criticized Benedetto Croce's thought (although the positive function performed by the neo-idealist philosopher).

The privileged *intellectual* target of *Prison Notebooks* is just Benedetto Croce. This is not surprising because, as Gramsci stated, to be an heir of the classic German philosophy means to be an heir of Crocean philosophy. But, at the same time, the cultural hegemony of Crocean neo-idealism had to be substituted by the philosophy of praxis, that Gramsci considered less speculative and closer to the masses. According to Gramsci, Proletariat is the new historical subject that moves forward to plead its own case just like in the Pellizza da Volpedo's painting *The Fourth State*.

Here the figure of the organic intellectual enters the scene, i.e. an intellectual that contributes to create a wide-reaching life, as Gramsci stated, an intellectual directly linked to the masses:

> The popular element "feels" but does not always know or understand; the intellectual element "knows" but does not always understand and in particular does not always feel. The two extremes are therefore pedantry and philistinism on the one hand and blind passion and sectarianism on the other. [. . .] The intellectual's error consists in believing that one can know without understanding and even more without feeling and being

impassioned (not only for knowledge in itself but also for the object of knowledge): in other words that the intellectual can be an intellectual (and not a pure pedant) if distinct and separate from the people-nation, that is, without feeling the elementary passions of the people, understanding them and therefore explaining and justifying them in the particular historical situation and connecting them dialectically to the laws of history and to a superior conception of the world, scientifically and coherently elaborated. [. . .] One cannot make politics-history without this passion, without this sentimental connection between intellectuals and people-nation. In the absence of such a nexus the relations between the intellectual and the people-nation are, or are reduced to, relationships of a purely bureaucratic and formal order. (Gramsci 2001, p. 1505)

Gramsci was firmly convinced that the intellectual and Italian moral reform could be only achieved thanks to his philosophy of praxis, i.e. the only *Weltanschauung* able to substitute Crocean hegemony also because, according to Gramsci, Benedetto Croce did not understand that the great intellectual must plunge himself into practical life. In short, Croce was not able to perform that function of cementing the formation of a new *historical bloc*, and his *Philosophy of Spirit* was also an antithesis of Gramscian philosophy of praxis (and also too abstract and speculative).

Historical bloc was just one of those fundamental steps for the realization of that *Weltanschauung* capable of overcoming the division between State and civil society existing in the coeval Italy:

If the relationship between intellectuals and people-nation, between the leaders and the led, the rulers and the ruled, is provided by an organic cohesion in which feeling-passion becomes understanding and thence knowledge (not mechanically but in a way that is alive), then and only then is the relationship one of representation. Only then can there take place an exchange of individual elements between the rulers and ruled, leaders and led, and can the shared life be realized which alone is a social force with the creation of the "historical bloc." (pp. 1505-1506)

Obviously, Gramsci realized the deep gap between East and West, consequently the essential point was to apply his theory of hegemony differently from the Leninian one because, as Gramsci writes:

In Russia the State was everything, civil society was primordial and gelatinous; in the West, there was a proper relation between State and civil society, and when the State trembled a sturdy structure of civil society was at once revealed. The State was only an outer ditch, behind which there stood a powerful system of fortresses and earthworks. (p. 866)

It means that the catch of power in the West cannot be concretized through a simple substitution of a leading class with one another (a similar decanting operation was doomed to failure). By contrast, hegemony had to find its development in civil society, that corresponds to "the function of 'hegemony' that a leading class exerts in all society," while political society, or State, absolves the function of a "direct domain or command that expresses itself in the State and in the 'legal' government" (pp. 1518-1519). Gramsci, in substance, puts civil society between the economic base and the State but, according to Gramsci, it is necessary to work inside civil society in order to realize the catch of power and to put the first bricks for the construction of hegemony, whose end should be the so-called *historical bloc*.

The concept of historical bloc is strictly linked to the thematic of the relationship between structure and super-structure that Gramsci conceives in a dialectical way and not on the base of a rigid cause-effect scheme.[3]

Gramsci does not consider ideologies as arbitrary just because, according to him, people get a sense of their social position just on the grounds of ideologies and here we can find a different way of conceiving the superstructure. In fact, according to Marx, the economic base is overriding in respect to superstructure while Gramsci considered superstructure as the privileged field where to work for the overcoming of Italian backwardness and provincialism. Gramsci, in this respect, seems to be a little bit Hegelian because he shows, on the one hand, that historical bloc has to manifest itself between economy and ideology, i.e. between economic structure and ideological superstructure, and, on the other hand, Gramsci provides full autonomy, and priority as well, to ideology (i.e. a super-structural element).

In his theory of hegemony, Gramsci takes into account the cultural and political factor. He thus establishes a dialectical relationship between structure and super-structure, which was not fully explicit in Marx. Most importantly, however, the hegemonic function is not a prerogative of an abstract social component, it pertains to the Political Party that identifies itself with the working class, which has to be aware of its historical role (see Maier and Semana 1978, p. 45).

---

[3] In effect, since his juvenile writings Gramsci understood the non-systematic nature of Marxian thought; so, according to him, for the study of Marxian theory, a critical and non-dogmatic approach was necessary: "Marx never wrote a little doctrine, he is not a Messiah who left us a string of parables gravid of categorical imperatives, of indisputable norms, absolute and out of space and time" (Gramsci 1975, p. 220).

Conclusions

As we have seen, Gramscian attempt at overcoming Italian provincialism was strictly linked to his philosophy of praxis, conceived as a *Weltanschauung*. In Gramsci's intentions, this was the only way of creating a national-popular culture capable of giving birth to a collective will. Italian cosmopolitism, whose notable representative was Benedetto Croce, was too provincial just because it did not create a national-popular culture.

Provincialism can manifest itself, generally speaking, in two ways: when a national culture is incapable of expressing or generating new ideas (lack of originality) or when a national culture is too close-minded, ignoring the ideas or cultures there are around the world.

Italian provincialism that Gramsci aimed to overcome was of the second type while provincialism in other countries, like Poland for instance, was of the first type. In Poland, for example, Leszek Nowak

> tried to prove that new interpretations of foreign philosophies have arisen in Poland but a really new philosophical trend has never arisen here. According to Nowak, great writers and essayists, not professional philosophers, were very close to it. The Polish mind is supposed to be provincial, held captive by the Western thought, and spiritually dependent on the Western Center.[4] (Górski 2007, p 149)

But was Gramscian attempt really successful in practice? Unfortunately, the answer is negative because the result of the organic identity between philosophy and politics forced the Italian Communist intellectuals that followed Gramscian lesson to act in the opposite way (as Friedrich Nietzsche would say, sometimes the cure is worse than the disease).

They tried to impose their ideology manifesting no sensibility toward other cultural currents, thereby paradoxically falling into the net of provincialism that Gramsci attributed to Benedetto Croce. In fact, on the one hand, Gramscian philosophy was hegemonic for a given period of time but, on the other hand, Communist intellectuals tried to isolate Italian culture from the most important philosophical ideas coming from foreign countries (just like Benedetto Croce did).

---

[4] Obviously, Leszek Nowak's thought is very far away from being a kind of provincialism, just because he gave birth to original epistemological and philosophical ideas. The real problem, in my opinion, finds its roots in the Polish language, which is snubbed by Anglo-Saxon culture.

The intellectual must be independent and he must exert his individual freedom of thought that, under the influence of any ideology, is impossible. Gramscian attempt at overcoming provincialism was full of good intentions but its inner structure was wrong because the organic intellectual must bend her individual freedom of thought to a specific political ideology.

Provincialism is thus always behind the scenes.

So, the essential problem lies in the relationship between hegemony and culture because when a given national culture wants to be nationally hegemonic it flows into provincialism. Then, what is the right way of overcoming provincialism? I think that, first of all, cosmopolitan and national elements must be kept in balance because a national culture cannot be isolated from the rest of the world. However, at the same time, a national culture must try to channel itself into the world cultural circuit, overcoming the narrow boundaries of its country.

Gramsci tried to overcome Italian provincialism but its problem-solving was not successful in practice because the organic intellectual, in his view, must necessarily bend his thought to political ideology.

The intellectual, in Gramscian terms, is obviously not a single individual (in this sense *we are all intellectuals*) but a social group who tries to be organic for a social class conceived as the carrier of the historical interests of the entire humanity (in this case this particular social class is the proletariat). But when an intellectual tries to be an *organic one*, he denies himself the chance of being a theoretical innovator (see Preve and Orso 2010).

Some of the intellectuals, as Leszek Nowak wrote,

> may support political power by constructing ideologies justifying the rulers, some may cooperate with revolutionary movements by elaborating utopias for them. In the first case they are representatives of the rulers, in the second, prophets of the people. But what matters is that in both cases their primary concern is to increase the number of believers in their ideology or utopia. (Nowak 1990, p. 170)

Following Leszek Nowak's distinction, we could say that Gramsci was both a prophet of the people (during Fascism) and a representative of the rulers after the fall of Mussolini's regime. But in both cases Gramscian *organic intellectual* did not provide any theoretical innovation because Gramsci's thought was exploited by the Italian Communist Party in order to justify its ideology.

The autonomy of reason, in Immanuel Kant's sense (*sapere aude!*), in this case cannot be applied just because all political ideologies force the intellectuals to think in a uniform way, preventing them to be *really critical.* The history of the most part of Italian intellectuals of the first half of the 20th century constitutes a valid proof of this anomaly.[5]

University of Catania
Dipartimento di Processi Formativi
Via Biblioteca, 4
95124 Catania
Italy
e-mail: giacomoborbone@yahoo.it

REFERENCES

Bobbio, N. (1977). *Gramsci e la concezione della società civile.* Milan: Feltrinelli.
Borbone, G. (2011a). The Legacy of Leszek Nowak. *Epistemologia* **34**, 227-252.
Borbone, G. (2011b). Per una nuova idea di scienza: Leszek Nowak e l'approccio idealizzazionale alla scienza. *L'arrivista (Quaderni democratici)* **3**, 34-48.
Borbone, G. (2012). Dialettica e filosofia della prassi: Marx filosofo idealista (in print). In: A. Medri (ed.), *Percorsi fondamentali dell'idealismo tedesco.* Villasanta: Limina Mentis Editore.
Cacciatore, G. (2005). Crisi e attualità del marxismo nel pensiero di Labriola. In: *Antonio Labriola in un altro secolo.* Soveria Mannelli: Rubbettino.
Coniglione, F. (1986). *Il sentiero interrotto.* Catania: Edizioni del Prisma.
Coniglione, F. (2008). Gramsci e il pensiero marxista tra storicismo e scienza. In: S. Salmeri and R. Pignato (eds.), *Gramsci e la formazione dell'uomo. Itinerari educativi per una cultura progressista.* Acireale-Rome: Bonanno.
Coniglione, F. (2010). *Realtà e astrazione. Scuola polacca ed epistemologia post-positivistica.* Second edition. Acireale-Rome: Bonanno.
Croce, B. (1993). *Storia d'Europa nel secolo decimonono.* Milan: Adelphi.
Fichte, J.G. ([1794] 1845). Grundlage der gesammten Wissenschaftslehre 1794. In: *Sämmtliche Werke,* vol. I. Berlin: Verlag von Veit und Comp.
Forgacs, D. (1999). National-Popular. Genealogy of a Concept. In S. During (ed.), *The Cultural Studies Reader.* London-New York: Routledge.
Górski, E. (2007). *Civil Society, Pluralism and Universalism. Polish Philosophical Studies,* vol. VIII. Washington: The Council for Research in Values and Philosophy.
Gramsci, A. (1917). La rivoluzione contro il "Capitale." In Gramsci (1973), pp. 130-133.
Gramsci, A. (1920). Operai e contadini, L'Ordine Nuovo, 3/01/1920, I, n. 32. In: Gramsci (1970), pp. 316-318.
Gramsci, A. (1966). *Socialismo e fascismo. L'Ordine Nuovo 1921-1922.* Turin: Einaudi.
Gramsci, A. (1970). *L'Ordine Nuovo 1919-1920.* Turin: Einaudi.
Gramsci, A. (1973). *Scritti politici,* vol. I. Rome: Editori Riuniti.

---

[5] Many thanks to Professors Krzysztof Brzechczyn and Katarzyna Paprzycka for their kind support, and to Professor Antonino Barbagallo and Professor Francesco Coniglione for teaching me the *relevance of culture.*

Gramsci, A. (1975). *Scritti giovanili 1914-1918*. Turin: Einaudi.

Gramsci, A. (2001). *Quaderni del carcere*. Critical edition of the Gramscian Institute edited by V. Gerratana. 4 vols. Turin: Einaudi.

Labriola, L. (1976). In memoria del Manifesto dei comunisti. In: L. Labriola, *Scritti filosofici e politici,* vol. II, ed. F. Sbarberi. Turin: Einaudi.

Maier, B. and P. Semana (1978). *Antonio Gramsci*. Florence: Le Monnier.

Marx, K. ([1845] 1997). Tesi su Feuerbach. In: *Scritti filosofici-giovanili*. Bergamo: Fabbri Editori. English translation in: F. Engels, *Ludwig Feuerbach and the Outcome of Classical German Philosophy* (New York: International Publishers, [1886] 1996).

Nowak, L. (1990). Intellectuals in the Age of Revolutions: The Case of the Socialist World. *Thesis Eleven* **27**, 167-172.

Nowak, L. (1992). The Idealizational Approach to Science: A Survey. In: J. Brzeziński and L. Nowak (eds.), *Idealization III: Approximation and Truth* (*Poznań Studies in the Philosophy of the Sciences and the Humanities*, vol. 25), pp. 9-63. Amsterdam/Atlanta, GA: Rodopi.

Pagano, M. (2007). Contemporary Italian Philosophy: The Confrontation between Religious and Secular Thought. In: S. Benso and B. Schroeder (eds.), *Contemporary Italian Philosophy. Crossing the Borders of Ethics, Politics, and Religion*. New York: SUNY Press.

Preve, C. and E. Orso (2010). *Nuovi signori e nuovi sudditi. Ipotesi sulla struttura di classe del capitalismo contemporaneo*. Pistoia: Editrice Petite Plaisance.

Rupert, M. (2005). Reading Gramsci in an Era of Globalising Capitalism. *Critical Review of International Social and Political Philosophy* **8** (4), 483-497.

Mieszko Ciesielski

HUMAN ON THE PERIPHERY OF COMMUNITY

WITOLD GOMBROWICZ ON PROVINCIALISM

ABSTRACT. In his literary works, Witold Gombrowicz has developed an interesting concept of a person entangled in a social sphere. A human being, according to the author of Ferdydurke, is an intrinsic being autonomically shaping his or her attitude in relations with other people. It is rather other people's circle, a social form, that fundamentally conditions the way a particular person thinks and acts. The depiction of an individual, portrayed by Gombrowicz, is a scale of attitudes ranging from the attitude of total submission to a social form to the attitude of peculiar freedom from the form. In my article I raise the questions of broadly understood "province-center" relations in the context of Gombrowicz's anthropological theses. Is the attitude of conformity and imitative adopting of thinking and behavior patterns characteristic of the "province" or of the "center"? Is the "province" doomed to imitative nature of thinking and coping of what the "center" does? Where is it easier to free oneself from the social influence and reach intellectual and cultural autonomy?

1. Introduction

The notion of "province" naturally refers us to the notion of the "capital," or preferably the "center." Both terms 'province'-'center' constitute a singular categorial system which reveals the dynamism of social reality they describe. Numerous scholars resorted to these notions as tools of adequate analysis of social phenomena. With their help, one attempted to grasp the mechanism of civilizational transformations consisting in bidirectional influence of the center on the province and of the province on the center. The difference of generally construed social potentials: economic, political, cultural ones etc. is supposed to cause flows between the two sides of the center-province system.

In social sciences, the interpretations of phenomena using categories of center-province or center-periphery were made by scholars adhering to the dependence theory, a school of the world system or post-colonial research.

In: Krzysztof Brzechczyn and Katarzyna Paprzycka (eds.), *Thinking about Provincialism in Thinking* (*Poznań Studies in the Philosophy of the Sciences and the Humanities*, vol. 100), pp. 103-119. Amsterdam/New York, NY: Rodopi, 2012.

The characteristic feature of these analyses is a strong partiality to a geographical understanding of the center-province opposition. Usually, the highly-developed European countries and the United States of America are considered to be the center while societies of the so-called Third World, countries of Africa and South America are taken to be the periphery.

The issue of province-center is also found in the work of Witold Gombrowicz.[1] Set against the strictly scientific analyses concerning the discussed opposition, the deliberations of one of the most prominent 20th century Polish writers appear to be very interesting, as Gombrowicz's perception of provincialism and centralism is not substantially related to the geo-economic or geopolitical distribution of societies. The categories of "province" and "center" are demarcated through particular features of interpersonal relations that the author describes in his vision of the human entangled in a social existence.

Issues of provincialism in Gombrowicz's work have been discussed on a number of occasions. For instance, Andrzej Falkiewicz emphasizes a peculiar immaturity found in Gombrowicz's writings, an immaturity in the sense of the indeterminacy of the provincial world, which seeks to boost its self-esteem through European culture.

> And he tracked down the indeterminacies of the world, of the others – the sloppiness of the immediate surroundings and the sloppiness of pre-war Poland, the cultural younger-being of the South American continent which clumsily strives to grow up to match the adult Europe. (Falkiewicz 1981, p. 58)

Leszek Nowak, in turn, relates Gombrowicz's deliberations on provincialism to the structure of scientific activity:

> Where is a province? The definition has nothing to do with geography: there, where one thinks to someone else's not one's own credit. Provincialism consists in a certain type of scientific career (especially in the humanities), in lacking the courage to throw the gauntlet at the world, but making a carpet of oneself, over which the "world" is to march to the Vistula or the Pripyat. (Nowak 2000, p. 191)[2]

In this paper, I will draw on Gombrowicz's understanding of provincialism in the light of a certain interpretation of his vision of the interpersonal human. Therefore, at the outset, I will interpret his concept of an individual embroiled in a "form," which will allow me to specify

---

[1] For an interesting biography of Witold Gombrowicz, see Siedlecka (2003).

[2] On provincialism in Gombrowicz's works, see e.g. Kępiński (2006), Jaszewska (2002).

three singular Gombrowiczian attitudes. Subsequently, these three attitudes will serve as an interpretation key to Gombrowicz's thoughts about provincialism.

2. Three Human Attitudes in Gombrowicz's View

In his writings, Witold Gombrowicz draws an image of a human "permeated with people." As we read in *Ferdydurke*:

> Indeed, in the world of the spirit a continual rape occurs, we are not self-contained, we are merely a function of other people, we have to be as we are seen. (Gombrowicz 1997c, p. 12)

Humans are not autonomous individuals who govern their lives. Their actions and attitudes are molded under the influence of the social surroundings. An individual is subjected to an unremitting tyranny of interpersonal behaviors, a tyranny of form born of human co-existence. A form is a structure that mediates human interactions. The author writes in *Ferdydurke*:

> the human being does not express itself directly and following its nature, but always in some specified form. (p. 79)

In the *Diary* one finds a telling fragment, a self-commentary of sorts:

> In my writings I have shown a man stretched on the Procrustean bed of form, I have found my own language to reveal his hunger of form and his dislike of it. [. . .] I have brought to light that sphere of the "interpersonal," which is decisive for people, and endowed it with the qualities of creative power. (Gombrowicz 1997a, p. 147)

The manner in which we are formed consists in that the values and goals which appear to be consciously accepted are in fact imposed by the community. We do not find valuable that which we consider to be such after consideration, but we are always somehow thrown into a world of values, into already established axiological hierarchies. This is how Leszek Nowak comments on this feature:

> It is not the individual who is the subject of preference, but the position within the form, and therefore in a way the social locus, which may be occupied by many different individuals. (Nowak 1996, p. 168)

Hence, the formation of an individual takes place by virtue of being imposed upon a specific hierarchy of values, which allows him or her to comprehend and arrange the surrounding world, and thus move efficiently within the social space. As Jan Błoński observes:

> For if there were no forms, or ready-made verbal, gesticular, moral codes
> and many others besides . . ., communing among people would be
> exceedingly hampered, slow, primitive. (Błoński 2003, p. 42)

Moreover, it needs to be noted that the Gombrowiczian form stipulates
the knowledge about the ways to actualize values. *Ferdydurke* offers a
description of punishment administered by Konstanty and Zygmunt to a
servant:

> never in the leg, never in the back, only walloped with their hands, clouted
> over, bashed the mug in! They did not fight with him – they did not beat
> him – just belted him on the kisser! And they were allowed to do so. It had
> been formally reserved for ages. (Gombrowicz 1997c, p. 253)

Thus the knowledge used in action is not objective and independent in its
content from human coexistence. Quite to the contrary, it is conditioned by
social relations. What we know and do not know is shaped by the social
form.

The vicissitudes of the protagonist of *Ferdydurke*, Józio, indicate that
an individual is not fatalistically doomed to being stuck in one form.
Gombrowicz presents a possibility of migrating from one form to another.[3]
With the proviso, and this is a significant feature of surmounting the form,
that exiting it must ensue in becoming simultaneously stuck in the
subsequent form. The author writes thus:

> there is no other escape from the mug, but into another mug, while from a
> human being one can only hide in the arms of another. (Gombrowicz
> 1997c, p. 264)

Hence the possibility of overcoming the form as such is rejected. The
only thing that individuals can bring themselves to do is to prevail over a
specific form, which means being imprisoned in another one. Gombrowicz
describes how the form is overcome by *Ferdydyrke*'s main protagonist on
two occasions. First, he manages to escape the form of "modernity," in
which the Młodziak family is bogged down. The second overcoming
comes with the destruction of the "country manor" form. Both in the
former and the latter case, the ultimate overcoming of the form consists in
realizing the fact of being molded by given forms. When analyzing the
concept of Gombrowiczian individual, Leszek Nowak writes thus:

---

[3] Overcoming forms, in the sense of surmounting barriers, transcending by Gombrowicz's
protagonists, is their basic trait in Edward Fiala's opinion, see Fiala (2002).

> Józio, cast into forms alien to him, acting in accordance with these forms, retains nevertheless his own "self" [. . .]. All the time, Józio is conscious of his duality. (Nowak 2000, p. 98)

This, in turn, causes the main character to "manipulate the form in which he was entangled, leading to its destruction" (Nowak 2000, p. 98).

Throughout her life, a person may overcome some forms and fall into others at the same time. The social life of an individual is constant maneuvering among forms. An exaggerated and grotesque description of precisely such existence may be found in *Ferdydurke*. Depending on the human approach toward the world of forms, three attitudes may be distinguished: that of the *formal human,* the *inter-formal human* and the *meta-formal human.*

The formal human. Such persons live all their lives in one and the same form. They never realize the fact of being imprisoned in a form, which would make it possible to overcome it. The attitude of the formal human is shaped by one and the same form throughout the entire life. Such persons do not destroy their forms, i.e. as long as they live they cultivate the same values, which they effectuate in the same fashion, employing unchanged specific knowledge. Simultaneously, the accepted values are treated as the only legitimate ones and as absolutely binding, values for which a fundament is sought in the transcendental reason. The knowledge possessed is considered true and infallible.[4]

The inter-formal human. This type of an individual is characterised by having overcome at least one form. Such a person mustered her strength to realize the conditions which shaped her life. This is a "first-degree" awareness of form, which takes the individual out of one form into another. The inter-formal attitude, despite succeeding in changing the form (once or more), also dogmatically accepts the values and knowledge imposed by the current form, taking them to be the only correct and true ones.[5]

---

[4] Such an attitude is found in *Ferdydurke* with the Młodziak family and the servants in Hurlecki's house.

[5] A description of the inter-formal approach may be found in a fragment of the *Diary* presenting the attitude of the communists: "as long as the destruction of the former truth is in question, that man [the communist] delights us with the freedom of the spirit of exposure, the desire of internal integrity; when however, seduced by the chant, we allow him to lead us to his own doctrine, bang! the door slams shut and where are we? Monastery? The military? A church? An organization? In vain may you try now to look for new addictions, which distort your new consciousness" (Gombrowicz 1997a, p. 307). The depicted communist may be interpreted as an inter-formal attitude; he has the knowledge of capitalism as a form, which is historical and relative with respect to economic interests.

The meta-formal human. This type of person attains the "second degree" of awareness of form, which represents a qualitative difference in comparison with the previous types. This consists in discovering and understanding the fact that people are continually imprisoned within a form. The meta-formal human is conscious that no set of values, worldview or ideology is absolutely true. Such a person recognizes that particular values, knowledge and social hierarchies stem exclusively from human coexistence and are not sanctioned by anything outside the social domain.

In *Ferdydurke*, Gombrowicz describes the life of the main protagonist, Józio, from that very standpoint. Toward the end of *Ferdydurke* he writes: "I'm running away with the mug in my hands." That "mug" is kept in hands, not in its place, i.e. on the head, because Gombrowicz is constantly aware of the human being permanently shaped by the forms. Holding the "mug" in hands means constant distance toward oneself as a person who is always in a form. Wondering whether *Ferdydurke*'s hero-narrator became "mature" enough for a specific form thanks to his ups and downs, Jan Błoński states:

> indeed, though not to just any form, but rather – to a dynamic attitude. The opposition of form and immaturity [. . .] cannot in fact be eliminated. Yet it is likely it can be outsmarted. Outsmarted with such conduct which will enable one to tempt another, and yet another man. . . in other words to amuse him, to arouse curiosity with one's own person, yet never letting oneself be determined, that is subjugated. (Błoński 2003, p. 65)

Still, it has to be stressed that that the meta-formal human is also stuck in a singular and unique form:

> How then, given the conditions, should one understand the struggle with the mug, the countenance in *Ferdydurke*? Clearly not in the sense that a man is to shed his mask – when he has no face beyond it – here one can only demand that he realizes his artificiality and confesses it. (Gombrowicz 1997b, p. 9)

The meta-formal person assumes a "dynamic approach," remains stuck in a special "meta-form" which imposes particular values, leading to non-involvement in religious, ideological or cultural debates. Let us quote the words of Gombrowicz which describe this attitude:

Still, the very same communist falls into a different form – the form of communism – of which he is not aware anymore. The values of another form, a form he is ignorant of, are the ones he takes to be absolutely true and dogmatically proclaims their superiority.

> Soon enough we will realize that this is not the most important: to die for
> ideas, styles, theses, slogans, beliefs; not this either: to be confirmed in
> those and shut oneself out: but this: to take a step back and achieve
> distance to all that ceaselessly happens behind us. (Gombrowicz 1997c,
> p. 83)

This singular "meta-form" enforces the attitude of distancing oneself
from oneself and the others, in a way which would make one an observer
of one's life and thereby experience one's own entanglement in the social
surroundings.

3. Provincialism according to Gombrowicz

The analysis of the issue of provincialism – provincial culture, provincial
thinking, provincial art, etc. – is carried out by Gombrowicz within the
framework of his vision of the interpersonal human, i.e. by means of the
notions of form and individual seeking complete self-determination within
the form, and struggling for his freedom from form.

Although in his descriptions of the province and the center, the author
of *Ferdydurke* resorts to examples of concrete nations – the countries of
Central Europe, especially Poland, are presented as the province while
Western Europe is shown as the center – this particular dimension of
deliberations is nonetheless not the crucial one. The Gombrowiczian
consideration attaches minor significance to the geographical location of
particular societies, whose historical determinants made them into cultural
centers of civilization or provincial margins. Each society, be it a
province, may equally well become a cultural capital, or even outperform
the center and make an original contribution to the global civilizational
development. The role that a given community is going to play is not
fatalistically conditioned by its history but in a way depends on the
strength of its reflection, on the courage of its intellectual elites to
undertake a mental coming to terms with the state of its own culture. In the
preface to *Trans-Atlantyk* Gombrowicz writes about the need to overcome
the constricting attachment to the nation:

> It would therefore mean a far-reaching revision of our attitude toward the
> nation. [. . .]. A revision, mark you, of universal nature – for I would
> suggest the same to the people of other nations, since the problem does not
> concern the attitude of a Pole toward Poland, but of a person toward the
> nation. (Gombrowicz 2005, p. 6)

Interesting depictions of the provincial attitude, as opposed to the attitude
typical of the center may be found for instance on the pages of *Diary*.

The fundamental trait of the provincial thought is its submissiveness with regard to the creators of the intellectual center. Artists, priests or scholars in the province relinquish independent thinking and adopt ready-made styles, norms and values developed by the center's cultural elites. Using the example of Poland, which Gombrowicz defines as a provincial country, the author describes the attitude of the province toward the highbrow culture:

> For a Pole, culture is not something he also contributes in creating, it comes from outside as something higher, superhuman – and he is impressed with it. (Gombrowicz 1997a, p. 245)

Yet this is not only the external culture, overwhelming with its grandeur and sacredness, the culture of the center to which one has to kowtow, that saps the spiritual powers of the provincial creator. The creator, or preferably a mere hack, crippled by the culture from over the way, by the grand culture, senses, more or less profoundly, the imitativeness, the smallness of his creation and tries to make amends for this state of subservience by a singular attitude of elevation. The provincial allows himself to be spiritually enslaved only to be able to hold sway over a still deeper province by the power of thought of the priests from the center.[6] The humility in the face of foreign thought is combined with the condescension in the face of the local provincial community. In a specific way, the province sees a fusion of the kneeling attitude with the one of being raised to the pedestal of worship.

The creators of the intellectual center do not experience the feeling that their expression lacks originality, a sensation that creator finds so humiliating. The culture of the center provides them with the comfort of authentic creative effort, allows them the style and content which are second to none, which are grand and fundamentally enrich an individual. When writing about the French, the English and the Americans, Gombrowicz states:

> they belonged to powerful and leading nations – to nations which not only did not ruin their personal lives but even enriched them – they were able to live to the fullest within the nation. (Gombrowicz 1997b, p. 24)

Conscious of the condition of Polish culture, construed as provincial thought, Gombrowicz puts forward a postulate of overcoming the label of a provincial:

---

[6] For a more detailed explanation of spiritual enslavement, see Nowak (2000), p. 179.

> To be nothing more in culture than a peasant, than a Pole, yet without overdoing it in being either. To be free but not to be too excessive in that freedom. (Gombrowicz 1997a, p. 146)

"without overdoing it" so as not to fall – let us add – into the gears of subsequent form, the "form of the free spirit," which, just as any other form, deprives a person of the awareness of being shaped.

When discussing the provincialism of culture, the author of the *Diary* devotes particular attention to the possibility of overcoming an opposition: the imitative rapture with the culture of the center versus negation of that culture and absolutization of the provincial norms. Here one perceives a singular program for the provincial creators, thanks to which they may go beyond the inhibiting center-province opposition. As Marcin Kępiński puts it in his study entitled "Gombrowicz's Games with Culture":

> Gombrowicz seems to be extraordinarily irked by the [. . .] vague attitude of the Pole toward the Western culture, which boils down to inferior imitation of its Polonized variant, or its negation and thus a shift toward the traditional national culture. (Kępiński 2006, p. 254)

The fundamental task that Gombrowicz sets himself is not the transformation of Polish provincial culture into a "wholesome" central culture. Such a change would still fail to radically overcome the determinacy and limitation of the spirit of provincial culture. Admittedly, Polish culture would fall from the clutches of imitative thinking into the embrace of a more original style, which nevertheless would still naïvely trust in the absolute, timeless and objective values. Gombrowicz desires a profound liberation, tries to evade the general categorization of "provincial" creator – "central" creator. Here is how the evasion might be accomplished:

> Yes! Be astute, judicious, mature, be an "artist," "thinker," a style-setter only to a degree while never being so too much, and make that "not too much" into strength equal to all very, very, very intense powers. (Gombrowicz 1997a, p. 146)

The way to avoid the imitative delight with the West and at the same time not to descend into unquestioning adoration of one's own national culture is a special reflective attitude which may cause the province to cease being yet another location where the concepts of the creators from the capital are copied. Reflective attitude is the approach capable of overcoming the center-province opposition, capable of elevating human spirit to the higher plane of social life:

> the only way in which I, a Pole, was able to become a fully-fledged phenomenon in culture is this: not to conceal my immaturity but admit it;

and by this admission to break away from it; and make a steed out of the
tiger that hitherto devoured me, a steed having mounted which I would ride
even further than those over there, the Western people, the "determined"
ones. (p. 263)

Gombrowicz is aware of the momentousness of his concept and
describes the originality of his vision of the interpersonal human, without
the complexes of a "provincial creator":

Perhaps more than many other authors, I have come closer in art to a
certain perception of the human – a human whose proper element is not
nature but people, a human not only situated within people but charged,
inspired with them. (pp. 147-148)

4. Province-Center and Gombrowiczian Attitudes

In light of Gombrowicz's remarks on the provincial society, let us consider
which of the three human attitudes is characteristic for the province and
which for the center? Which human attitude: the formal, the inter-formal
or the meta-formal creates the culture of the center, and which is typical of
the province?

The basic feature of the cultural center is the originality of thinking. In
this sense, the creators of the cultural center devise ethical, aesthetic and
epistemic norms, which shape the lives of the inhabitants of the center as
well as of the province, which imitates the values and thinking of the
central culture. Every now and then, the artists, scholars and priests of the
central societies abandon the dominant style, criticize the state of science,
the norms in force and suggest alternative means of artistic expression,
scientific concepts or systems of values. The center repeatedly overcomes
the form, to use Gombrowiczian terminology, in which it functions.

The inter-formal human is therefore the typical attitude for the center.
They can achieve the first degree of the awareness of form and overcome
the form they have been in so far, through which they have expressed
themselves and which shaped their relationships with the other members of
the center. The spiritual life of the center is formed by norms, values and
theories created by artists, priests and scholars originating from that very
center. At this point, a singular originality of the central thought is
revealed, which autonomously breeds new visions and systems by seeking
new styles of expression, new ethics, new scientific concepts etc.

The provincial societies may generally be divided into unreflective and
reflective provinces. The division stems from the dominant attitude
adopted by its people. In the unreflective province, one encounters only

formal or inter-formal attitudes. Meanwhile, the reflective province owes its nature to the meta-formal attitude.

The creators of the unreflective province follow the path set by the thought leaders from the center. The formal and inter-formal attitudes are characteristic for this province. We may therefore distinguish two varieties of the unreflective province: the active and the passive one. The active provincial thought follows the center all the time. Whenever the creators of the center change their form of expression into another, so the provincial creators change theirs in emulation. Thus the active province, in that it permanently duplicates the center, is typified by the attitude of the inter-formal human.

Next to the active province we can discern the unreflective passive province, where the creators retain a form once adopted from the center for a long time and do not change it, despite the fact that the center introduces further changes in its thinking. The formal human is the attitude characteristic of the passive province. The once imitatively adopted form (religion, morals, art style) is preserved uncritically, as it is judged as the only true and correct one.

The difference between the center and the unreflective province consists in the fact that the center dresses up in a form, which is new and fresh, whereas the person of the province attires himself in an already worn, second-hand form which has been slavishly taken from the center. He faces reality in this tried form. The tested form adopted from the cultural center is safe indeed but it does not let one discover new truths concerning reality and humans.

In his description of the provincial thought, Gombrowicz gets the better of the widespread belief that originality and innovativeness are the sole competence of the central culture, while the province is doomed to derivative, imitative participation in human culture. This holds true for the unreflective province, where the inter-formal and formal attitudes predominate. Next to it, one finds the reflective province, which in its characteristics departs both from the unreflective province and the societies of the center.

Specifically, a province becomes reflective thanks to its meta-formal attitude. The attitude attains the second degree of awareness of the form, i.e. possesses the knowledge that any action (be it artistic, religious, moral, political or economic) results form social relationships, from the interpersonal tension that enforces particular behaviors on our part. The meta-formal human is conscious of the relative nature of values and norms whose validity stems from interpersonal coexistence, not from some absolute reason, which is transcendent with respect to the society. The knowledge enables one to assume a certain attitude of non-commitment. It does not allow social forms to take their ultimate shape but it lets one maintain the distance toward oneself, other people and reality.

A creator from the reflective province overcomes the stereotypical understanding of the center-province opposition. Despite its originality in searching for more styles, values and truths, the center is not capable of self-reflection and attaining the fundamental truth, viz. that the justification and binding force of successive stages of human culture springs from the social life of the people. The truth about the human being as the sole force, which brings forth the styles, values and religions, may be grasped by one who displays the meta-formal attitude, the attitude characteristic for the reflective province. One should not adhere to or believe in norms and values too strongly – neither the original, newly-discovered ones (as is done by the priests of the center), nor the tried imitated ones (as is done by the provincial unreflective thought).

Could we point to any example of a reflective province? According to Gombrowicz, Poland could become such a reflective province. How could Provincial Poland become a reflective province or a fully-fledged participant and a force that creates universal culture, shared by all humankind? Gombrowicz answers:

> From a Pole who is proud of himself, boasting about himself, enamored of himself, make a being who is most acutely conscious of his/her insufficiency and temporariness – and make that sharpness of vision, the ruthlessness in not concealing the weaknesses into a strength [. . .] our entire attitude toward the world would have to change, while it would not be our duty anymore to develop a specific Polish form, but to effect a new approach to form as something which keeps being created by people, and yet always fails to satisfy. More than that: it has to be demonstrated that everyone is like us, that is to unmask the whole insufficiency of the civilized human with respect to culture, which is beyond him. This is about nothing more nor less than a transposition from a man having a form into a man (this also applies to nation) producing a form – a dry recipe perhaps, but one which suddenly and unexpectedly changes the entire Polish manner of being in the world. (pp. 263-264)

Krzysztof Stala, commenting on Gombrowicz's program of surmounting the inflexible, long-established Polish form, the romantic, Catholic, patriotic one, writes thus:

> By rejecting the official Pole, the official, mature Polish culture, Gombrowicz discovers a Pole who is alternative, immature, who cold-shoulders not only Polish myths and stereotypes but form in general. This lax attitude to form, the anti-idolatrous approach to culture, Europe, the world – that precisely is the chance for a Polish intellectual. (Stala 2002, p. 150)

Gombrowicz confronts the attitude of the Western individual with the attitude that might be born in Poland. The successive stages of Western culture, the successive styles in art, the successive beliefs and the paradigms in science were each time finally recognized as correct and true, requiring to be fully identified with. The reflective provincialism of Gombrowicz is not aimed at particular religion or ideology but at that naïve faith in the finality of an axiological system and excessive attachment thereto:

> Imagine the shock – when the proud "I am French" of the Frenchman and "I am English" of an Englishman" encounters the unexpected Polish "I am not quite Polish, I am above Polish." (Gombrowicz 1997b, p. 72)

Gombrowicz outlines a vision where provincial Poland outstrips a great nation, the culture of the center. The higher level of spiritual life, the original attitude of a Pole is to rely on a singular personal approach to culture, to form as such:

> My proposal to Poles was an attitude toward the nation that was more radical – and without precedent – something that would set us apart from the mass of nations, made us a nation with a different and unique style. (p. 24)

Poland, just as any other country with provincial culture, might become a culturally dominant center, one which sets new ways of thinking about the human being in the context of his/her entanglement in culture, nationality and other forms of social existence. That new way of thinking consists in realizing that it is the human who is the ultimate authority legitimizing values and norms, not some god, history, reason of state, class interest etc. However, the human here is not construed as an individual being, with potent subjectivity and conviction of her absolute freedom, self-steerability and rationality of her actions but as an interpersonal being, embroiled in a permanent process of self-creation among people, at times jointly with them, at times in spite of them.

> What will be born, what *could* be born in Poland and in the souls of ruined and brutalized people, when one day even the new order which smothered the old one disappears, and Nothing ensues? Here is a picture: a stately edifice of a thousand-year old civilization fell, it is empty and still, while on the rubble pile – a swarm of gray and small human figures, who still cannot shake off the bewilderment. For there collapsed the church, the altars, the paintings, the stained glass and statues before which they knelt [. . .]. Where to find shelter? What to adore? Whom to pray to? Whom to fear? [. . .] Would that be strange if they saw in themselves the only force that creates and the only Deity accessible to them? This is the path which

leads from adoration of the products of man do the discovery of man, as the decisive and naked might. (Gombrowicz 1997a, p. 36)

Gombrowicz's postulate to vanquish provincial thinking may be expressed as the imperative to develop a meta-formal attitude within oneself. The attitude evinces a higher level of social awareness, which consists in the recognition of the belief in social, interpersonal sources of artistic, religious, scientific and other values. The awareness of the temporariness as opposed to the finality of styles, systems and axiological systems, enables one to take a different look at the human being. The new perception of the individual, his non-self-containedness and the entanglement in interpersonal existence enables the human to be determined as the single authentic source of value. Such a perspective on the human, and thus the attainment of the meta-formal attitude, requires that he be accepted in his naked existence, without the intermediate garb: religious belief, nationality, citizenship, the economic role of the consumer/producer etc. Dominique de Roux formulates the concept thus:

> The words "above all, the human should one day meet the human" is most certainly a fundamental maxim. Its analysis reveals a thesis which it is appropriate to denote a "Gombrowiczian philosophy." (Gombrowicz 1996, p. 10)

Can a Pole adopt the meta-formal attitude? Gombrowicz responds in the negative:

> The inhabitants of the magnificent edifice of the Western civilization should be prepared for the invasion of the homeless, with their new feeling for the man. . . an invasion that will not come to pass [. . .] For a Bulgarian does not trust a Bulgarian, A Bulgarian despises a Bulgarian. (Gombrowicz 1997a, p. 36)

The Bulgarian – symbolizing European province to which Poland also belongs – will not attain a new cultural perspective, new spirituality, as he is too attached to expressing feelings for the fellow human via intermediate bodies: the patriotic love of the nation enables brotherhood with the compatriots, the sacred love of God enables Christian love of one's neighbor, the desire for profit enables respect for a business partner. In Polish province, just as on any other, one sees an entire spectrum of feelings mediated by patriotism, religion, the functioning of the market. Only the common confidence in a human as a human is found wanting.

Provincialism in Gombrowicz's vision may be surmounted thanks to the meta-formal attitude, which means resignation from the imitative adoption of norms, values, styles, which compels to criticism in thinking and courage in individual evaluation. A reflective creator from the province is capable of distancing herself from the center-province arrangement and consequently of demonstrating its limitations. This presents an attempt to overcome this opposition. Such an attempt will not usually enjoy wide acclaim since, as a rule, an original provincial creator is pushed to the margins of her own community. This is because the visions and programs she puts forward do not gather following. They are rejected for the basic reason that the lineage of their creator is too native, too provincial. On top of that, the autonomous, center-independent provincial thought lays utterly bare the very inconvenient truth: the submissiveness of the provincials to the culture of the "center":

> If, among the provincials, there appears someone who has the courage to compete for the values with the center, not only does he fail to receive help but is also simply driven down by his compatriots. The enslaved do not dare to compete with the master by definition, so they hate those of their kinsmen who muster the courage for spiritual rebellion. For the latter show the former who they really are through their sheer existence. (Nowak 2000, p. 179)

It is much easier to preach slogans: "We must catch up with Europe," "We must become Europeans at last," which in practice means passive and unreflective adoption of the cultural, political and economic models from the countries of Western Europe. It is more difficult to call "Let our Europeanism be a trial run," "Make Europe to human measure, not vice versa." The attitude was aptly described by István Eörsi in the book *My time with Gombrowicz*:

> Let us also speak the word "market," but let our enthusiasm stop short of orgasm. Let us say "Western democracy," but let us not forget that even that creation does not guarantee survival of the humankind. (Eörsi 2005, p. 47)

Gombrowicz's case illustrates the lack of confidence in native creators and the associated inferiority complex perfectly. The author of *Ferdydurke* was aware that his vision transcends far beyond the modern understanding of the human as a self-steerable being, endowed with freedom and autonomy in deciding about their own life. The beginnings of systematic development of the concept of a rational human in Western Europe met with criticism uttered by a Polish writer, a provincial writer. The criticism was not limited to a negative rejection of the Western philosophy of the human, but suggested a new concept of the human entity – a human who has the awareness of the relative nature of values, of the inability to develop an objective point of view, a singular temporariness and the sense of life being a specific game, where the lack of final rules is not a hindrance but, on the contrary, allows one to ascend to a higher plane of sensitivity. The extent to which Gombrowicz's vision is similar to the post-modern perception of the human and society is convincingly demonstrated by Dagmara Jaszewska:

> In *Ferdydurke,* there lingers a spirit of the post-modern era, whose ideas were to evolve only half a century later. Although Gombrowicz lived in 1904-1969, and *Ferdydurke* was written in the interwar period, he may be considered a poet of post-modernity, an excellent analyst and expert on the condition of the human contemporary to us.[7] (Jaszewska 2002, p. 9)

Witold Gombrowicz desired to make the Polish province a reflective one, which would outdo the naïve Western concept of autonomous and rational human, and which would simultaneously overcome the communist idea of the social human being, propagated in the post-war Eastern Europe. However, the voice of the author of *Ferdydurke,* coming from the periphery of the community, was not audible enough and did not contribute in the development of humanist thought in a way it might have done.[8] And today, amid the loud cries of "Catch up with Europe" coming from various provincial corners, can we hear the contemporary Gombrowiczs, creators who are original, one of a kind, who put forward bold visions of the human, society and the world, which clash with the "binding" and "politically correct" concepts?

Adam Mickiewicz University
Institute of European Culture
ul. Kostrzewskiego 5-7
62-200 Gniezno
Poland
e-mail: mieszko@amu.edu.pl

REFERENCES

Błoński, J. (2003). *Forma, śmiech, i rzeczy ostateczne. Studia o Gombrowiczu* [Form, Laughter and Things Final]. Kraków: Universitas.

---

[7] On the links between Gombrowicz's thought and postmodernism, see also: Szahaj (1996), Pieszak (1999), Ciesielski (2006).

[8] Naturally, one encounters contrary views, which assert minor significance of Gombrowicz's vision, see e.g. Kępiński (1994).

Ciesielski, M. (2006). Witolda Gombrowicza wizja człowieka a jednostka epoki ponowoczesnej [Witold Gombrowicz's Vision of Man and the Individual of the Post-Modern Era]. *Zeszyty Filozoficzne* **12-13**, 223-233.

Eörsi, I. (2005). *Mój czas z Gombrowiczem* [My time with Gombrowicz]. Warszawa: Instytut Badań Literackich PAN.

Falkiewicz, A. (1981). *Polski Kosmos* [The Polish Cosmos]. Kraków: Wydawnictwo Literackie.

Fiała, E. (2002). *Homo transcendens w świecie Gombrowicza* [Homo Transcendens in Gombrowicz's World]. Lublin: Wydawnictwo KUL.

Gombrowicz, W. (1996). Testament. Rozmowy z Dominique de Roux [The Testament. Conversations with Dominique de Roux]. Kraków: Wydawnictwo Literackie.

Gombrowicz, W. (1997a). *Dziennik 1953-1956* [Diary 1953-1956]. Kraków: Wydawnictwo Literackie.

Gombrowicz, W. (1997b). *Dziennik 1957-1961* [Diary 1957-1961]. Kraków: Wydawnictwo Literackie.

Gombrowicz, W. (1997c). *Ferdydurke*. Kraków: Wydawnictwo Literackie.

Gombrowicz, W. (2005). *Trans-Atlantyk*. Kraków: Mediasat Poland.

Jaszewska, D. (2002). *Nasza niedojrzała kultura. Postmodernizm inspirowany Gombrowiczem* [Our Immature Culture. Postmodernism Inspired by Gombrowicz]. Warszawa: Oficyna Naukowa.

Kępiński, M. (2006). *Gombrowicza gry z Kulturą* [Gombrowicz's Games with Culture]. Warszawa: Wydawnictwa Akademickie i Profesjonalne.

Kępiński, T. (1994). *Witold Gombrowicz Studium portretowe II* [Witold Gombrowicz, Portrait Study II]. Warszawa: Alfa Wydawnictwo.

Nowak, L. (1996). Gombrowicza model świadomości (między)ludzkiej [Gombrowicz's Model of Inter(human) Consciousness]. In: A. Falkiewicz, L. Nowak (eds.), *Przestrzenie świadomości. Studia z filozofii literatury* [The Realms of Consciousness. Studies in Philosophy of Literature] (*Poznańskie Studia z Filozofii Humanistyki*, vol. 16), pp.138-192. Poznań: Zysk i S-ka.

Nowak, L. (2000). *Gombrowicz. Człowiek wobec ludzi* [Gombrowicz. The Human versus the People]. Warszawa: Prószyński i S-ka.

Pieszak, E. (1999). Gombrowicza nierzeczywista rzeczywistość [Gombrowicz's Unrealistic Reality]. *Opcje* **3** (26), 6-13.

Siedlecka, J. (2003). *Jaśnie Panicz. O Witoldzie Gombrowiczu* [The Honourable Young Master. On Witold Gombrowicz]. Warszawa: Prószyński i S-ka.

Stala, K. (2002). Gombrowicz – wobec Polski, wobec Europy [Gombrowicz – and Poland, and Europe]. In: D. Kolano (ed.), *Szukanie głosu Gombrowicza* [Looking for Gombrowicz's Voice], pp. 146-161. Radom: Teatr Powszechny im. Jana Kochanowskiego w Radomiu.

Szahaj, A. (1996). *Ironia i miłość. Neopragmatyzm Richarda Rorty'ego w kontekście sporu o postmodernizm* [Irony and Love. Richard Rorty's Neopragmatism in the Context of the Debate on Postmodernism]. Wrocław: Lepoldinum.

PART III

PROVINCES AND CENTERS OF THE WORLD

Adolfo García de la Sienra
Leandro Rodríguez Medina

HISPANIC-AMERICAN PHILOSOPHY

IN THE FRINGES OF THE EMPIRE

ABSTRACT. After presenting a brief history of philosophy in Hispanic-America since the XVI century, we discuss whether the idea of province and empire is applicable to contemporary Hispanic-American philosophy, investigate the form these ideas adopt in this region, and inquire into the ways in which provincialism is present in philosophical work. We conclude that there are three main groups that understand their peripheral position in different ways, with different views on the way in which they should insert (or not) into the mainstream of Western philosophy.

1. Introduction

The aim of the present paper is to investigate whether the idea of province and empire is applicable to contemporary Hispanic-American philosophy, to investigate the form these ideas adopt in this region, to inquire into the ways in which provincialism is present (and what forms it takes) in philosophical work. We will also discuss the roles that provincial philosophers are doomed (or not) to play. We believe that the awareness of the issue is helpful in intellectual proceedings and that Hispanic-American philosophers have special duties nowadays.

The idea of province and empire, at least in the realm of science and philosophy, has to do with leadership and dependency. As Alatas (2003) has pointed out:

> In addition to considering the role of social scientific research and scholarship in the service of political and economic imperialism, we may also think of it as analogous to political and economic imperialism, that is, the "domination of one people by another in their world of thinking." In other words, academic imperialism is a phenomenon analogous to political economic imperialism. There are imperialistic relations in the world of the

In: Krzysztof Brzechczyn and Katarzyna Paprzycka (eds.), *Thinking about Provincialism in Thinking* (*Poznań Studies in the Philosophy of the Sciences and the Humanities*, vol. 100), pp. 123-139. Amsterdam/New York, NY: Rodopi, 2012.

social sciences that parallel those in the world of international political economy. Academic imperialism in this sense began in the colonial period with the setting up and direct control of schools, universities and publishing houses by the colonial powers in the colonies. It is for this reason that it is accurate to say that the "political and economic structure of imperialism generated a parallel structure in the way of thinking of the subjugated people." These parallels include the six main traits of exploitation, tutelage, conformity, secondary role of dominated intellectuals and scholars, rationalization of the civilizing mission, and the inferior talent of scholars from the home country specializing in studies of the colony. (Alatas 2003, p. 601)

Can Hispanic-American philosophy, or at least part of it, be characterized as provincial in this sense?

In order to characterize a school or individual as central or peripheral/ marginal, we propose to look at the strategies of those individuals or groups making philosophy in the periphery. We can divide the individuals into three groups: (1) those that belong, on equal footing, to a metropolitan center, even if they live or spend most of their time in Hispanic America; (2) those that accept a peripheral status but nonetheless try to participate, in a significant way, in the dialogue of central schools; (3) those that accept their peripheral status as a position from which it is not necessary or relevant to establish a dialogue with metropolitan centers or nets. The strategies adopted can be called *integration*, *insertion*, and *self-exclusion*, respectively.

Those that follow the exclusion strategy usually adopt some philosophy that originated in a metropolitan center, like Germany or England, and seek to "apply" this philosophy to formulate or address local problems (like "the identity of the Mexican," for instance). Yet, they never try to publish in the top metropolitan journals where the integrated members of the school usually publish, or to participate in the international conferences in which the latter participate. Typically, they do not speak the language of the school either.

In the first part of this paper we will present a (necessarily brief) history of Hispanic-American philosophy, starting with Scholasticism. We shall see that Scholastic philosophy in American lands was a vigorous, authentic movement, fully integrated with the Spanish academic institutions (and the Catholic Church). Outstanding among the Hispanic-American philosophers were the Jesuits, disciples of Francisco Suárez and Luis de Molina. We shall see that their bold eclectic stance, which originally tried to strike a synthesis between the religious motive of humanism (autonomy of the will) with Christian religion, ended up creating a mood that made the attempt to synthesize any school of philosophy with Christianity easy and natural. This mood gave rise to the Eclectics in the eighteenth century, a new Jesuitic attempt that tried to synthesize philosophy with Christian faith. In the twentieth century, and beginnings of the current one, this mood has been present especially in "liberation philosophy." We shall see that after a century of political turmoil, the nineteenth, in which academic philosophy was hardly cultivated (with some exceptions), and which finished with the adoption of positivist philosophy in almost all the countries of the region, the different schools of European philosophy at the beginnings of the twentieth century made their appearance in Hispanic America. At the beginning of the twentieth century there appeared in Hispanic America a generation of self-taught philosophers who had great social influence. Called the "founders" by some (incorrectly, since the real founders of Hispanic-American philosophy were the Scholastics, as we shall see), they discredited positivism and set the stage for the reception of other philosophical trends. Yet, the introduction of serious scholar philosophy in the Hispanic-American universities was due to the second generation of twentieth-century philosophers, called the philosophers of the normalization. Almost all these philosophers were followers of Husserl or Heidegger. This division is crucial to the understanding of Hispanic-American philosophy at large. Husserlian philosophers were concerned with the problem of the foundations of knowledge and their movement eventually introduced analytical thinking in Hispanic America. Heideggerian philosophers found in Heidegger's ontology the tools to approach the problems of Latin America and eventually gave rise to Latina-Americanist thinking and "liberation philosophy." The first bifurcation was staged by the third generation, which arose around the middle of the 1960s. Those philosophers that had been concerned with the "problem of knowledge," with phenomenological epoché, and the question whether there could be found apodeictic foundations for knowledge, discovered the rigor of analytical philosophy and became fond of authors like Carnap (who visited the Mexican analytical community several times toward the end of the 1960s), Frege, Russell or Strawson. Those who had been more inclined to Heideggerian ontology were more concerned with the "identity" of the Latin American. Some others believed that Marxism would provide the solution to what they considered the main problem of Hispanic America: a form of dependent capitalism that provoked poverty and social inequality. The third generation only accentuated the bifurcation distancing even further the analytical philosophers from the Latin-Americanist and the Marxist traditions.

In the second part, we shall discuss the question whether any of these two movements (which still characterize Hispanic-American philosophy up to now) are provincial or not, what is their position vis à vis "the Empire"; i.e. the great philosophical centers of the West; and what are the obligations of philosophers in Hispanic America.

2. Brief History

2.1. *The Beginnings*

Hispanic America is a conglomerate of countries that extends from the US southern border down to Patagonia, including some Caribbean islands as well. These countries are the children of Spain and the prevailing language in all of them is Spanish (although in some of them, like Mexico, there exist millions who still speak only some native language). Having been born as colonies of Spain, the philosophy that was cultivated in them after the Conquest, and up to the eighteenth century, was not original from these lands, but was introduced into them by the religious orders. Notoriously, the first school of philosophy in the American continent, founded by Fray Alonso de la Vera Cruz in Michoacan, Mexico, was created to educate the native nobility in practical arts, but also in philosophy, as Fray Alonso taught there his philosophy course during several years since 1540. His philosophical works are deemed as the first works on philosophy produced and published in the New World.[1] Indeed, up until the eighteenth century, Late Scholasticism was the predominant, if not the only philosophical stream in Mexico and the rest of Hispanic America.[2]

All the schools and universities in Hispanic America were dominated by the Scholastic way of thinking, thus reflecting the complete dominion of the Catholic Church in education.

To begin with, was the introduction of Scholastic Spanish philosophy in the New World a form of academic imperialism? The question sounds rhetorical but it is more difficult than it might seem at first sight since it can hardly be said that the setting up of the new Spanish academic structure "generated a parallel structure in the way of thinking of the subjugated people" (Alatas 2003, p. 601). The structure was not "parallel," because philosophy properly speaking was not cultivated in these lands before the arrival of the Spanish friars. At any rate, all previous academic

---

[1] The first one was the *Recognitio Summularum* (Mexico, 1554), followed by *Dialectica resolutio* (Mexico, 1554) and *Phisica Speculatio* (Mexico, 1557). Fray Alonso was a nominalist in logic.
[2] An extraordinary, thorough bibliography of this philosophical production is found in Redmond (1972).

structures were completely obliterated by the conquerors. Leaving aside the lower classes, which were illiterate anyway, the upper classes – whether Indian, Spanish or mixed – certainly came to be educated within the framework of Catholicism and Scholasticism. This was unlike the British rule in India, which intended to bring an ancient culture closer to the British, grafting in some cases its institutions onto existing indigenous structures (cf. Baber 2003, p. 616). Hence, as a matter of fact, the academic structure introduced by the Spaniards in their American colonies was going to be the historic foundation of any such future structure. Much of the struggle of the liberals toward the end of the eighteenth century, and all of the nineteenth, was indeed to destroy Catholic education and replace it by a "modern" one; i.e. German or French, so that this process of modernization was not the rise of a national (or Latin American) consciousness but rather the search for a new foreign tradition from which local intellectual projects could be derived.

Scholastic philosophy was intensely cultivated in Hispanic America until, at least, the second half of the eighteenth century, when the winds of modernity began to blow in the viceroyalties and the Jesuits were expelled (1767), in spite of the watchful eye of the Inquisition. The question is whether the schools of philosophy in the New World were provincial or were at a par with, say the University of Salamanca.

Were the Hispanic-American Scholastic scholars peripheral or marginal? Were they recognized as peers on equal footing by their colleagues at Salamanca (the main philosophical center of Europe in the sixteenth century) and other Peninsular centers (like Coimbra)? The semiotic-material network of Scholastic thinking in Hispanic America was built by the Catholic Church through its religious orders, mainly the Jesuits and the Dominicans, and so we must assume that there was a certain continuity in the way the colleges and universities were established in America and Spain. On the other hand, unlike other colonial powers, Spain never saw its colonies with contempt but tried to reproduce itself with splendor in them, mainly in New Spain (Mexico). As a matter of fact, it has been said that Fray Alonso was the most brilliant student in Salamanca, apparent heir to the greatest figures, like Francisco de Vitoria or Domingo Báñez, when he decided to move to the New World.

Thus, Spanish colonization intended to generate a "continuation," a recreation of Spain in the New World. The intellectual production in Hispanic America was conceived as metropolitan as that of Salamanca, especially after the consolidation of the academic institutions. The transfer of the standards and criteria of intellectual production was direct and not the result of a semi-critical appropriation of foreign parameters (as usually takes place between centers and peripheries). Moreover, the physical displacement of the philosophers from the Peninsula to America did not mean degradation, but rather the assumption of a civilizing mission that in many ways represented an aggrandizement of the Empire.

The productivity of the American philosophers was outstanding and in certain cases (like that of Tomás de Mercado) they were an important reference for their peers in Spain. It is beyond doubt, however, that no philosopher in America ever reached the heights of Francisco Suárez or Luis de Molina, even though there were very brilliant exponents of Suárez's philosophy, like Diego Marín de Alcázar in New Spain. As a matter of fact, as Kuri (2008) has shown, the philosophical production of the Jesuits in New Spain was interesting, vibrant and with a certain air of originality; i.e. they developed in a scientific and systematic way the implications of Suárez's and Molina's views, in full accord with the worldview and religious drive animating them, had the material conditions to develop and circulate their ideas, and were recognized by other Jesuits as interesting and relevant. One Dominican that was widely recognized in Spain was Tomás de Mercado, who wrote an important logic textbook and a treatise on contracts that is considered an important work on Scholastic economics.[3]

As evidenced in several publications (for an almost complete list, see Redmond and Beuchot 1985), there were quite a few outstanding (terminist) logicians in Hispanic America, among which we have mentioned already Fray Alonso and Mercado. Another important figure was Antonio Rubio. Overall, we can say that even though the philosophers of the colonial period in Hispanic America were not entirely original, and that the centers in the New World were never as important as Salamanca or Coimbra, nevertheless they maintained at least the same intellectual standards as their Peninsular peers, were engaged in the same disputes and problems, and so they had a certain authenticity. They were integrated to the Spanish network of philosophers and eventually they even produced original arguments and books of standard quality. Their contribution to educate the Americans (indians, creoles and *mestizos*) in the dominant Spanish worldview is beyond question. In that sense, they were true creators of civilization and culture.

---

[3] *Comentari lucidissimii textum petri Hispani*, Seville, 1571. *Suma de tratos y contratos*, Seville, 1567.

2.2. *The Reception of Modern Philosophy*

The Peace of Westfalia consecrated the religious division of Europe marking the rise of the religious motive of secular humanism – autonomy of the will. The theological disputes, so common until the seventeenth century, began to be replaced by rationalist philosophical discussions. Fueled by the impressive success of Newtonian mechanics, all forms of Enlightenment (the Scottish Enlightenment, the *Aufklärung* in the German territories, or the more aggressive French Illustration) gained the upper hand in Europe. Even in Spain, the stronghold of Catholicism, important changes took place, mainly in the public administration, with the arrival of the Bourbon kings after the War of Succession (1701-1714). The Bourbons began to displace the Baroque Jesuit worldview with a sort of rationalist perspective, mainly in statecraft and architecture, a displacement that reached its peak in 1776 with the expelling of the Jesuits from all the Spanish dominions.

Even though, according to Oliver (1998), the expelling of the Jesuits "delayed the introduction of proto-modern European philosophy in Latin America," the process had started more than one century before 1776 since, in Hispanic America, modern philosophy was received in the middle of the seventeenth century in the guise of eclecticism or "elective philosophy," an intellectual movement that finds its zenith in the beginning of the nineteenth century. Great eclectic philosophers were Juan Benito Díaz de Gamarra and Francisco Javier Clavijero, in México; José Agustín Caballero and Félix Varela, in Cuba; Francisco Javier de Santa Cruz y Espejo, in Ecuador; Cayetano Rodríguez and Elías del Carmen Pereyra, in Argentina. Eclectic philosophy was used by these authors as a method to introduce the new scientific and philosophic theories: Cartesianism, the French Encyclopedists, and scientific experimental methods. This led to an educational reform in the universities of Hispanic America, where even authors banned by the Inquisition were read. Eclecticism gave to the Hispanic-American intelligentsia a new common identity and a new synthesis between (Catholic) faith, experience and reason. This new view was going to be the foundation of the "independentist movement," as it began to exalt the human virtues, the native Americans (the indians), the creoles, and the American fatherland vis-à-vis Spain. It meant a serious rift with the Spanish dominant, authoritarian, scholastic views. According to Moreno (1973, p. 201), "the very reform of the curricula and the new idea of philosophy have the end of creating in the American man an intelligence that, not deserving anymore the label of barbarian, enable him to reach terrenal happiness." That may be right but, as we shall see, eclecticism, the joyful blending of views that seem hardly compatible, became a dominant attitude and trait of Hispanic-American intellectuals.

The real problem for the development of philosophy along the nineteenth century in Hispanic America was the political turmoil both in various Hispanic-American countries and in Europe. Universities occasionally closed. This inhibited academic philosophical progress as universities are the natural locus of philosophical activity. A more productive forum for philosophy was often the political arena in which thoughtful essays were written by non-academics on themes such as constitutional government, progress and autonomy (cf. Oliver 1998). Later in the nineteenth century, positivism became eventually entrenched on most Iberian-American countries. For instance, in Mexico, President Benito Juárez had sent Gabino Barreda to France, to study the doctrine at the feet of Auguste Comte. The effects of the positivist education on the ruling classes were powerful in the reorganization of public administration in Mexico, Brazil, Argentina and other countries (Brazil's flag still maintains the positivistic dictum: *ordem e progresso*, i.e. order and progress). In Argentina, the positivist worldview shaped museums, universities, schools, public health as well as public administration and political organization. In Mexico the thinking of the *científicos*, the organic intellectuals of the Porfirio Díaz regime, was thoroughly positivistic.

2.3. *The Twentieth Century*

The twentieth century in the philosophical arena of Hispanic America started with a generalized rejection of positivism. Hispanic-American thinkers accused positivism of scientist and began to explore many other possibilities. At the beginning of the century, in opposition to positivism, Hispanic-American philosophers "entertained idealism, vitalism, pragmatism and various political and social philosophies. Neo-Thomist thought continued to be widely studied, primarily in the Catholic universities" (Oliver 1998), a trend that is now as strong as ever. These new trends established Hispanic-American academic universitarian philosophy. These first Hispanic-American twentieth-century philosophers have been dubbed "the founders" but this designation is unfair since the founders were the scholastics in the colonial era. We can keep the term, nevertheless, if we understand that there was a certain lack of continuity between the colonial and the independent academia, which was mainly due to the turmoil (already mentioned) that took place along the nineteenth century but also to an explicit rejection of that tradition by modern Hispanic-Americans.

Following Miró Quesada (1974) and Dussel (2011), we may divide the history of Hispanic-American philosophy in the twentieth-century into four stages, which we will discuss in turn.

2.3.1. *The Founders*

The most notorious among these were the Colombians Alejandro Deustua and Carlos Arturo Torres, the Uruguayan Carlos Vaz Ferreira, the Argentineans Alejandro Korn and Coriolano Alberini. In Mexico, after the 1910 Revolution, Antonio Caso and José Vaconcelos were outstanding intellectual figures. According to Dussel,

> This generation of "founders" has great relevance. They were philosophers who thought about their reality with conceptual tools that they forged, frequently autodidactically, and who ought to be rediscovered for contemporary reflection. (Dussel 2011, p. 16)

Miró Quesada (1974, pp. 30*ff*) said of them that their consciousness was sundered by a "lack of focus," since they reflected a philosophy for which reality was European. This is nevertheless a strange criticism: How can reality be European for philosophy – for *any* philosophy? Is not philosophy a *Wissenschaft* and, as such, entirely universal?

We shall see that this is the axis around which the disputes about the character of Hispanic-American philosophy revolve. The dilemma is this: should Hispanic-American philosophy be universal, insert itself in the mainstream of world (i.e. Western) philosophy as did Scholasticism in the colonial era? Or should it be rather devoted to the development (or adaptation) of a conceptual apparatus in terms of which the "identity" of Hispanic America and Latin America or its social problems would be considered?

2.3.2. *The Second Generation or the Normalization*

The second generation of Hispanic-American philosophers is characterized by their serious scholarly character. An outstanding educator and founder of institutions was the Marxist Argentinean educator and politician Aníbal Ponce (who died in Mexico in 1938). Danilo Cruz Vélez is perhaps the most important Colombian philosopher of the twentieth century. As a pupil of Heidegger in Freiburg, he became a very important expert in his philosophy and vindicated, in consonance with Francisco Romero, the right of the Hispanic-Americans to be a deliberating part in the wider context of Western philosophy. Thus Cruz Vélez as well as Romero can be seen as an advocate of the strategy of integration. Other Heideggerian scholars, more inclined to "use" Heideggerian ontology to reflect about

Latin America, were the Spaniard José Gaos in Mexico, Carlos Astrada in Argentina, Alberto Wagner de Reyna in Peru, Félix Schwartzmann in Chile and Ernesto Mayz Vallenilla in Venezuela. There were others, like the Spanish-born Argentinean Francisco Romero, who were followers of Husserl, Hartman or Scheller. The "professionalization" of philosophy had an important impact on the intellectual production of this generation. They developed and established a set of criteria to evaluate intellectual production, such as writing structures, publications' guidelines and so on. In 1949 Romero coined the term 'philosophical normality' in order to refer to the exercise of philosophy as an "ordinary function of culture in Hispanic America."

2.3.3. *The Third Generation. The First Bifurcation*

The rift between the "cosmopolitans" and the "localists" (devoted to a so-called "Latin American philosophy") became more evident in the third generation. The following statement, uttered by Juan Bautista Alberdi in a speech he gave in Montevideo in 1842, has all the romantic and historicist ingredients that gave rise to Latin-American philosophy.

> The philosophy of each epoch and each country has commonly been the reason, the principle, or the most dominant and general sentiment that has governed its actions and conduct. And this reason has emanated from the most imperious needs of each epoch and country. Thus there have been Greek, Roman, German, English, and French philosophies, and it is necessary that there be an American philosophy. [. . .] There is, then, no philosophy in this century; there are only systems of philosophy, that is, more or less partial attempts, contradictory among themselves. (Quoted by Marquínez Argote 1981, pp. 12-13)

Hence, an authentic Latin American philosophy should be something like the self-explanation of a supposed Latin American *Volksgeist*.

The most outstanding Latin-Americanist is undoubtedly the Mexican Leopoldo Zea who, starting as a historian of ideas, attempted to answer the question "What is our being?" and ended up with an anti-Western stance. According to Miró Quesada, Zea's philosophy became involved in

> a more intense and harder struggle because Western domination encounters allies in our peoples, in groups of oligarchical power who also speak of freedom but only to defend their interests, interests that coincide with those of the foreign dominators. [. . .] Zea moves from the philosophy of what is Mexican to what is [Latin] American, and then, in a stage of maturity, to the philosophy of the Third World. [. . .] This humanist integration of humanity and its history is, today, the horizon from which unfold the

theories of cultures of dependence and in which the philosophy of liberation has many roots. (Miró Quesada 1981, pp. 149, 183)

Other roots of the "philosophy of liberation" are to be found in Marxism.

2.3.4. *The Second Bifurcation*

The Second Bifurcation was marked by the rise of analytical philosophy in Hispanic America and signaled an even deeper chasm between the Latin-Americanists and the "subjectivists." Many of the former phenomenologists, like Francisco Miró Quesada and Mario Bunge motivated by the methods of symbolic logic and the logical analysis of language, moved toward some form of analytical philosophy. According to Dussel:

> All were influenced by the "linguistic turn" in postwar Anglo-Saxon thought. Philosophy in Latin America took a clear forward step – although the project of quasi-perfect rigor of mathematical formalization or analysis meant an exaggerated skepticism regarding the other currents, and by the eighties, the limits of its internal and external consistency, especially with respect to practical philosophy, had been discovered. In any event, Latin American philosophy became conscious of its own methodological-linguistic mediations. (Dussel 2011, p. 27)

The analytical movement gave rise to very serious academic institutions in Hispanic America, like the *Instituto de Investigaciones Filosóficas* (IIF-UNAM) of the National University of Mexico and the *Sociedad Argentina de Análisis Filosófico* (SADAF). The founding father of IIF-UNAM was Eduardo García Máynez, an outstanding philosopher of law but the institution was consolidated by Fernando Salmerón, Alejandro Rossi and Luis Villoro, who also created the journal *Crítica: Revista Hispanoamericana de Filosofía*. The journal was born with the pretense of being as international and cosmopolitan as any good European or US-American journal. SADAF was founded by eminent Argentinean figures like Carlos Alchourón, Eugenio Bulygin, Genaro Carrió, Alberto Coffa, Juan Carlos D'Alessio, Ricardo Gómez, Gregorio Klimovsky, Raúl Orayen, Eduardo Rabossi, Félix Schuster and Thomas Moro Simpson toward the end of the 1960s. In 1981, SADAF founded *Análisis Filosófico*, a journal that intends to be no less international and serious than *Crítica*. The great avatars of the philosophy of science in Hispanic America are Gregorio Klimovsky, Mario Bunge and C. Ulises Moulines. They introduced the most strict standards of rigor and a fully international, non-provincial, Western mentality into the Hispanic-American philosophy of science. The structuralist group in the philosophy of science has given rise to another international journal, *Metatheoria*, published in Buenos Aires.

Dussel suggests that philosophy of liberation started with his work Para *una ética de la liberación latinoamericana* in 1971, after which

> a group of philosophers emerged at the Second National Congress of Philosophy (Cordoba, 1972), whose discourse had as its point of departure the massive poverty of the underdeveloped and dependent Latin American continent. (Dussel 2011, p. 33)

The "liberation philosophers" thought that, in order to be rigorous and authentic, philosophy had to depart from the concrete and elevate to universality. Against Leopoldo Zea, Augusto Salazar Bondy believed that a new philosophy was required with the attributes of being

> Illuminating with respect to the question of "negativity," and more linked to praxis in the question of social "transformation." To achieve this, the social sciences, the political economy of dependence (today we would say the horizon of the "world-system"), must be assimilated. (Dussel 2011, p. 32)

Thus, clearly, the "liberation philosophers" are thinking of liberation as liberation from the capitalist "world-system" (a term coined by Immanuel Wallerstein to refer to the current global economic system), which is deemed by them as the culprit of the "oppression" of the "Latin American peoples." But, as García de la Sienra (2007) has shown, the religious motive that dominates Wallerstein's (or Marx's for that matter) worldview is that of autonomy of the will – and there is hardly a motive more Occidental than this.[4] Hence, the liberation philosophers are found in the awkward position of being under the grip and control of one the defining motives of Western culture in order to formulate a non-Western philosophy! Religious motives are universal, in the sense that they make take hold of any nation, tongue or "ethnic group," but the Freedom-Nature motive, in point of actual fact, has taken roots only in the West.

3. Assessment and Expectations

As we have shown in the first part of this paper, the origins of Hispanic-American philosophy lie in the Scholastic tradition that Spanish philosophers brought to the new continent. Thus, our brief account challenges contemporary history, according to which the founding fathers

---

[4] Cf. McCarthy (1990) for a fully detailed explanation of the form this motive adopts in Marx. For the concept of religious motive, see Dooyewerd (1979), chapter 1.

of Hispanic-American philosophy were philosophers who lived and produced toward the end of the nineteenth century, and who introduced some of the main debates from European philosophy. Our short history also demonstrates that the development of Hispanic-American philosophy is associated with ruptures, i.e. "bifurcations," that have been taking place during the last century. Such bifurcations consolidated a professional philosophy in the region. However, at the same time, they illustrate the extent to which Hispanic-American philosophers have been in debt to European traditions such as Husserlian phenomenology or Heideggerian ontology.

Interestingly enough, the bifurcations that grouped philosophers around three major traditions (subjectivist, analytical, and Marxist or liberationist) share similar characteristics. First of all, they received their main intellectual influence from metropolitan scholars and institutions, such as Carnap or the Frankfurt School. Second, they attempted to professionalize local philosophy by developing curricula at universities and launching local publications that could appropriate (and sometimes expand) European debates. Third, they were subjected to the interference of the state into the intellectual life, in different ways, throughout the twentieth Century. From *coups d'etat* in the South Cone to organic intellectuals in Mexico, Latin American philosophers of all traditions have had to fight for the autonomy of their field, sometimes giving in to the interest of state's authorities. Fourth, the material resources for a professional philosophy have always been scarce. On one hand, philosophy has been practiced mainly in public institutions and, as such, has been dependent on public funds that are usually cut off when economic crises arise. On the other hand, the lack of private funding has affected the whole education system, making it difficult to find a private institution with a long and serious tradition of philosophical reflection. In addition, scarce resources have been an expelling force for many scholars who have had to leave to find better working and living conditions. Finally, the traditions that emerged in Hispanic-American philosophy do not always recognize the importance of the plurality of thought in the development of the discipline. More often than not, they formed sects with parochial interests that have not attempted to produce a sustainable and rich dialogue. These sects, with different vocabularies, methodologies and research interests, end up breaking up the field, impeding the consolidation, if possible, of a Hispanic-American philosophical tradition.

These similarities, however, must not obscure the relevant differences between them. First of all, each group has followed a different strategy to participate in the (international) philosophical debates. While subjectivists and Marxists have opted for insertion or self-exclusion, analytical philosophers usually chose integration. The reasons could lie in their understanding of being in a peripheral context. For analytical philosophers, philosophy is a kind of universal reflection that can be made from any location, is based on reasoning and, for that reason, anyone can make a contribution. For this group, contribution means the participation in the international debate, i.e. publishing in mainstream journals and establishing such journals in peripheral contexts (e.g. *Crítica* or *Metatheoria*).

Subjectivists and Marxists (including "liberation" philosophers) have understood their position in the international arena in terms of the belief that the loci of enunciation affect the final product of reflection. Put differently, according to them philosophy is embedded in particular contexts from which some ideas can emerge while others are silenced, and so philosophical reflection is not universal but situational. Paradoxically, these thinkers (always under the grip of the Freedom-Nature motive) have used metropolitan philosophers as inspiration (e.g. Husserl or Habermas), have participated in the international debate (e.g. the work of Dussel has been translated into English, French and other imperial languages), and have attempted to adapt foreign theories to the local (Latin American) context. In this situation, the use of the locus of enunciation as an epistemological foundation of their philosophy seems to be an instrument to exclude other participants in their debates. Only those who live (and suffer) in Latin America are valid interlocutors. Why? Because context matters. How? It seems to shape the socio-mental frameworks within which understanding is possible. However, if this is the situation, how is it possible for peripheral philosophers to understand Heidegger's philosophy properly? How is it possible for them to bring Derrida's *différance* to America? Is the context not important when translating European theories into Hispanic-American philosophical reflection? This is the surface of the deep inconsistency that burdens the localist philosophers.

When insertion and self-exclusion are the strategies, the cost of being peripherally situated increases. For "inserted" scholars, the peripheral position is usually used as a way to excuse them for not being able or willing to fully participate in international debates. We take this to be quite a provincial stance indeed. Frequently, they claim that the lack of resources prevents them from developing original, rigorous scholarly activities and, as a consequence, they do not meet the standards of mainstream journals or international conferences. For "self-excluded" scholars, the peripheral situation seems to be the perfect alibi to embrace mediocre canons. They argue that periphery is not only a structural configuration of the international division of academic and intellectual labor but also that it is a way of seeing a specific location from which some problems can be approached and eventually solved. For them, there are peripheral standards that should regulate the production of local knowledge but they are not seen as epistemologically problematic or as sources of new ideas. This is why some of the works of peripheral scholars end up being an apology of poverty.

Insertion and self-exclusion can be nevertheless valid, though costly strategies. Many Australian and Finnish academics, for instance, think of themselves as "inserted" scholars because they understand their position in the international scenario but attempt to take advantage of it. Peripheral conditions need not affect the quality of contributions if some socio-political and economic conditions are met (e.g. full-time professorships or up-to-date libraries). Self-exclusion, on the other hand, would require scholars to ignore the metropolitan theories (and standards of scholarship) altogether. If the locus of enunciation is as important as many of them really think then only those who share that situation (a kind of *Lebenswelt* or *Weltanschauung*) are valid participants. Excluding however, is always risky, since it is also a way of silencing voices that, although different, could challenge previous understanding.

Two of the most important conclusions of this article are that (a) there is no homogeneity in the periphery and (b) there are many ways of being a peripheral philosopher. Periphery is not a simple category to describe non-metropolitan fields, institutions and academics. Peripheries share some characteristics (e.g. the lack of institutionalization or resources) but have important differences too (e.g. languages and intellectual tradition). These differences open up possibilities for scholars to experience the peripheral condition diversely, which leads to integration, insertion, and self-exclusion as the main strategies. In any case, philosophers in peripheral contexts should not forget or ignore their role in their societies, because "peripheral" is a contemptuous concept only when it means resignation and uncritical acceptance of foreign standards of knowing – and living. In particular, Hispanic-American academic philosophy should recognize its Western origins and roots and adopt the standards of scholarship that are internationally accepted.

Universidad Veracruzana
Instituto de Filosofía
Tuxpan 29
Fracc. Veracruz
91020 Xalapa, Veracruz
México
e-mail: asienrag@gmail.com

Universidad de las Américas Puebla
Departamento de Relaciones Internacionales y Ciencia Política
Ex Hacienda de Sta. Catarina Mártir S/N
San Andrés Cholula, 72820, Puebla
México
e-mail: leandro.rodriguez@udlap.mx

REFERENCES

Alatas, S.F. (2003). Academic Dependency and the Global Division of Labour in the Social Sciences. Current Sociology 51 (6), 599-613.

Baber, Z. (2003). Provincial Universalism: The Landscape of Knowledge Production in an Era of Globalization. Current Sociology 51 (6), 615-623.

Dooyeweerd, H. (1979). Roots of Western Culture. Toronto: Wedge Publishing Foundation.

Dussel, E. (1970-1975). Para una ética de la liberación latinoamericana. Buenos Aires: Siglo Veintiuno Argentina.

Dussel, E. (2011). Philosophy in Latin America in the Twentieth Century: Problems and Currents. Retrieved December 26, 2011 from http://www.afyl.org/dussel1.pdf.

Ferrater Mora, J. (1953). Suárez and Modern Philosophy. Journal of the History of Ideas 14 (4), 528-547.

García de la Sienra, A. (2007). ¿Desarrollo o revolución? Estado, Economía y Hacienda Pública 13, 49-54.

Kuri, R. (2008). El barroco jesuíta novohispano: la forja de un México possible. Xalapa: Universidad Veracruzana.

Marquinez Argote, G. (1981). ¿Filosofía latinoamericana? Bogotá: El Búho.

Miró Quesada, F. (1974). Despertar y proyecto del filosofar latinoamericano. México: Fondo de Cultura Económica.

Miró Quesada, F. (1981). Proyecto y realización del filosofar latinoamericano. México: Fondo de Cultura Económica.

McCarthy, G.E. (1990). Marx and the Ancients. Savage: Rowman & Littlefield.

Oliver, A.A. (1998). Philosophy in Latin America. In: E. Craig (ed.), Routledge Encyclopedia of Philosophy. London: Routledge. Retrieved December 22, 2011, from http://www.rep.routledge.com/article/ZA009.

Redmond, W.B. (1972). Bibliography of the Philosophy in the Iberian Colonies of America. The Hague: Martinus-Nijhoff.

Redmond, W.B. and M. Beuchot (1985). La lógica mexicana en el Siglo de Oro. México: UNAM.

HISTORICAL REFERENCES

Fray Alonso de la Veracruz, 1554. Recognitio, Summularum Reverendi Patris Illdephonsi a Veracruce Agustiniani Artium ac sacrae Theologiae Doctoris apud indorum inclytam Mexicum Primarii in Academia Theologiae moderatoris. México: Imprenta de Juan Pablos.

Fray Alonso de la Veracruz, 1554. *Dialectica resolutio cum textu Aristotelis edita per Reverendum Patrem Alphonsum a Vera Cruce Agustiniani m. Artium atq, sacrae Theologiae magistrum in achademia Mexicana in nova Hispania cathedrae primae in Theologiae moderatorem.* México: Imprenta de Juan Pablos.

Fray Alonso de la Veracruz, 1557. *Phisica Speculatio, aedita per R. P. F. Alphonsum a Veracruce Agustinianae familia Provintialem, artium, et sacrae Theologia Doctorem, atque cathedrae prima in Academia Mexicana in nova Hispania moderatorem.* México: Imprenta de Juan Pablos.

Tomás de Mercado, 1571. *Comentarii lucidissimi in textum Petri Hispani*, Fernandi Diaz, Seville. Spanish translation: *Comentarios lucidísimos al texto de Pedro Hispano* (México: UNAM, 1985).

Tomás de Mercado, 1571. *Suma de tratos y contratos*, Seville. Recent edition: (Madrid: Instituto de Estudios Fiscales, Ministerio de Economía y Hacienda, 1977).

Patryk Pleskot

DOES HISTORIOGRAPHY NEED TO BE PROVINCIAL?

INTERNATIONAL CIRCULATION OF IDEAS AS EXEMPLIFIED BY THE COOPERATION OF POLISH AND FRENCH HISTORIANS IN THE PERIOD OF THE PEOPLE'S REPUBLIC OF POLAND

ABSTRACT. Contacts between Polish historians, French historians and French centers of historiography – especially with the prestigious milieu of Fernand Braudel's *Annales* – were unusual and extraordinary in comparison with other forms of scientific cooperation with foreign countries: both with the West and with the "friendly countries." Because of the undeniable uniqueness of these relations many scholars from various countries claim that the annalistic methodology "influenced" Polish historiography. What is characteristic, however, is that these statements are most often completely a priori. This paper is a reflection on the nature of the methodological influence of one historical school on the other and discusses such a possibility, taking into consideration models of circulation of ideas proposed by Pierre Bourdieu and Jerzy Maternicki. It is also an attempt at answering whether historical sciences are able to freely interfere on a supra-national level or whether they are by nature characterized by provincialism, understood here as a limitation to national frameworks outside of which they cannot be understood.

Contacts between Polish historians, French historian and French centers of historiography – especially with the prestigious Parisian milieu of Fernand Braudel's *Annales* – were unusual and extraordinary in comparison with other forms of scientific cooperation with foreign countries: both with the West and with the "friendly countries." Despite the Iron Curtain, every year from the late-1950s until 1989 a few dozen Polish scholars paid research visits to Section VI of EPHE/EHESS,[1] some of whom established

---

[1] VI Section of EPHE – École Pratique des Hautes Études was established in 1947. In 1975 the Section transformed into a fully-independent École des Hautes Études en Sciences Sociales – EHESS. It was often referred to as just School – École. Together with prestigious

In: Krzysztof Brzechczyn and Katarzyna Paprzycka (eds.), *Thinking about Provincialism in Thinking* (*Poznań Studies in the Philosophy of the Sciences and the Humanities*, vol. 100), pp. 141-154. Amsterdam/New York, NY: Rodopi, 2012.

close relationships or friendships with their French counterparts. It was the only form of cooperation between scholars from these two countries to reach such significant proportions (see Pleskot 2010). Because of the undeniable uniqueness of these relations, many scholars from various countries claim that the annalistic methodology "influenced" Polish historiography. What is characteristic, however, is that these statements are most often completely *a priori*. This paper is a reflection on the nature of the methodological influence of one historical school on the other and discusses such a possibility. It is also an attempt at answering the question whether historical sciences are able to freely interfere on a supra-national level or are they by nature characterized by provincialism, understood here as a limitation to national frameworks outside of which they cannot be understood.

A clear and exhaustive answer to these complicated questions is situated on the borderline of historiosophy and the sociology of science. The work of the sociologist Pierre Bourdieu (2002), who is only now being discovered in Poland, might be the key to the analysis of these issues. In his reflections on the international circulation of ideas (or to use the original term – "intellectual import-export"), he analyzes the obstacles in the way of the international circulation of ideas, including research methods. According to Bourdieu:

> it is often thought that intellectual life is spontaneously international. It cannot be further from the truth. Intellectual life is a place, as any other social space, in which there are nationalisms and imperialisms. Intellectuals, like other people, transmit stereotypes, prejudices, simplifications, very general and summary ideas stemming from every-day events, misunderstandings. (Bourdieu 2002, p. 3)

Thus "international exchange is subject to a certain number of structural factors which imply misunderstandings" (p. 4). What factors are these? The most important is that texts circulate outside the context in which they were written (Bourdieu focuses on the circulation of publications and translations), i.e., to use Bourdieu's expression, outside of their mother context of the "scientific field" or "production field" (understood as "activity field," or "sphere"). This might occasion misunderstanding of the author's intentions in the country receiving a given work; especially because the receivers are in the "reception field" of that text, which implies reinterpretation or misunderstanding. Hence, "the meaning and function of a given foreign work are determined at least to the same extent

---

periodical *Annales* edited by the School employees it made up the milieu known as the Annales School.

by both the mother field and the reception field" (p. 4). Moreover, the lesser the familiarity with the conditions in which a given idea was created, the more probable the deformation of its original thought (p. 7).

Bourdieu points out other traps in the international circulation of ideas. The differences between the "mother field" and "reception field," usually corresponding with the "national field (and for us with provincialism) might, according to Bourdieu, produce "false discrepancies between similar things and false similarities between disparate things" (p. 6). This issue is connected with another phenomenon: receivers of a foreign thought take into consideration not what the author says but what is imputed. That is why it is mostly "flexible" scholars who win international recognition because their ideas might be interpreted in a number of ways (p. 5). It should be noted that the *Annales* historiography could be considered a "flexible" paradigm – one that allows for absorption of the elements of Marxism, structuralism, sociology, anthropology, etc. The Annales School was so capacious a notion that many scholars of varying opinions, methodologies, but perhaps not poetics (as understood by Philippe Carrard), could subscribe to it.

It should be noted that Bourdieu's reflections concern almost exclusively French-German relations. As in the case of Polish-German relations, barriers to intellectual understanding grew as a result of a tradition of mutual animosity, past and present-day conflicts, political and cultural-civilizational rivalry, taboo subjects, etc. As for Polish-French relations, most of these factors hampering the circulation of methodological ideas could have been less intensive. Moreover, the international circulation of ideas does not occur exclusively through translated texts, which Bourdieu focuses on, but also by means of reading original publications as well as thanks to direct meetings (conferences, conversations, discussions) or indirect ones (correspondence). On these levels of exchange, some of the French sociologist's reservations lose significance.

So could Polish historians have yielded to the Annalists' influences and have stepped out of the provincial frame in their methodology? To explain these issues, it seems necessary to define 'methodological influence'. The term is rooted in psychology, anthropology and the sociology of science or the history of ideas. It seems indispensible to differentiate between "influence" and "inspiration." Krzysztof Pomian attempted to perform such a classification by differentiating between "receptivity" and "penetration," though he did not define the term (Pomian 1978, p. 119). In this article, I assume that we can talk about "influence" when, as a result of a conversation with a French colleague or reading his book or attending a lecture, etc., a Polish scholar makes a methodological or technical decision, which he could not have made without that conversation/book/lecture, for example, when he changes his research interests or approaches a given topic anew. "Influence" is the most visible evidence of transgression of provincialism.

We can talk about "inspiration" when that conversation/book/lecture changes to some extent (modifies or partly directs) the Polish scholar's technical-methodological convictions, whose basic core remains unchanged. Here we could talk, for example, about the completion of an already developed conception, noticing its new aspects and similarity in opinions, leading to a discussion based on partnership, etc. The issue of provincialism is less obvious here. If the scholar basically maintained his views then did he become part of the international circulation of ideas? The phenomenon which I call "intellectual mimesis" would be even lower in that hierarchy of dependencies. In this case both scholars' visions are similar and known to the other party, but mutual inspirations or influences cannot be proved. These similarities might stem from the fact that these two scholars belong to the same intellectual community, the same "hermeneutic circle." But if these two scholars had similar thoughts, even though they did not have to know each other, then they simply could not have been stuck within provincial boundaries.

"Knowledge" – Polish historians' awareness of the Annalists' achievements – is the basis and *condicio sine qua non* for both "influence" and "inspiration." It should be noted that knowledge does not prove a transgression of provincialism. At the same time, without knowledge there can be neither inspiration nor influence, but on the other hand knowledge does not automatically imply these two phenomena. It should be added that "mimesis" does not stipulate "knowledge," but is actually more easily noticeable without it. Naturally, in many cases it is very difficult to separate "familiarity" from "inspiration" or "influence." For example, does a reference to a given historian in a footnote or quotation prove only "knowledge" of him or is it evidence of "inspiration" or perhaps of "influence"? It seems that in order to fully answer this question we would have to analyze each work of a given author separately, which is impossible. Secondly, does polemics with a given methodology (for example Peter Burke's 1978 description of English historians' allegations against the Annalists) provide evidence of "familiarity," because you can not criticize something that you do not know about, or is it perhaps already an evidence of "inspiration" (polemics is a manifestation of one's view on a given subject matter and one's interest in it; while interest implies given research decisions – even if that very critique!).

On the basis of a broad search query[2] and what I read I would venture to assume that in the People's Republic of Poland there was good "knowledge" of the *Annales* milieu achievements from before the war and from the Braudel period at least since the 1970s.

Andrzej Wierzbicki's (2004) findings indirectly confirm this thesis. Not only all of the most influential Polish historians of historiography (Marian Henryk Serejski, Andrzej Feliks Grabski, Jerzy Topolski, etc.) but also a significant number of other scholars who specialized in other fields were able to provide a detailed list of the achievements of the scholars from Lucien Febvre and Fernand Braudel's circle. Practically each of the École visiting researchers, and there were hundreds of them, had an idea (from a vague to a clear one) about the institution owing to their visits to Paris. But what about "inspiration"? Its existence seems undeniable and obvious, even though it concerns a circle of scholars narrower than the circle of those who manifested the "knowledge." The very fact of staying in Paris, conversing with French colleagues and attending seminars or choosing certain books determined a smaller or larger dose of "inspiration." As for "influence," I would suggest that it is practically impossible to prove its existence using the typical technical methods of a historian.

It should be pointed out that intellectual influence cannot be likened to "technical" influence. What does this distinction mean? "Technical influence" might be characterized based on the example of the work plan of Zygmunt Gostkowski – a VI Section intern in 1965. He stated that one of the aims of his stay in France was to "become familiar with modern methods of archiving ... sociological data and with organization of comparative sociological research on an international scale" (Gostkowski 1965, no pagination). So Gostkowski's adoption of these techniques would be an example of "technical influence," for in that case it was not about intellectual ideas but about the "craft" connected with the "operation" of knowledge. Archivists' and teaching system specialists' visits to France had a similar character as they were at the borderline of science and technique. Unfortunately, it is not usually as easy to isolate "technical influence" in its pure form as in Gostkowski's case. "Technical influence" often blends with "methodological influence" in an unfathomable way. It is obvious that during a VI EPHE Section stay in Paris the teaching system practiced in that school had a direct influence on Polish scholars. The

---

[2] While working on the (2010) book devoted to the Polish historians' relations with the *Annales* milieu during 1945–1989, I did research in 25 Polish and French archives and institutions.

influence was less or more significant (sometimes it could have resulted in discontent with the Section teaching system), however, it did exist simply because the scholars learned about the functioning of the School as well as of libraries and archives. However, it did not have to mean that the scholar had to use the new methodology in his own work. He could simply become familiar with it from the perspective of craft.

The analysis of the *Annales* methodology's influence on Polish historians gets even more complicated as the influence was both direct and indirect. If we make a theoretical assumption that, for example, Marian Małowist yielded to the influence of Fernand Braudel and others, then we might suspect that he in turn influenced his students. So can we say that those students were inspired by French historiography? The answer is complicated. Michał Tymowski stressed the fact that his article with which he arrived in Paris within 1970 was immediately printed in *Annales* as it corresponded with the periodical's profile.[3] Could it have resulted from his intellectual formation as Małowist's student? It is probable with a reservation that Małowist himself was inspired by the French achievements. In turn, another student of the professor, Maria Bogucka, recalls that it was her stay in France and not the Polish seminars that broadened her scientific horizons by the "Annalistic" areas.[4] We cannot, obviously, entirely trust witness testimonies – in that case they could have been shaped by, for example, the nature of the relations with Professor Małowist.

Alfred Dubuc discusses such problems in his analysis of *Annales'* influences in Quebec. The scholar wonders if in Canada it is possible to differentiate between the presence of Henri Berr's concepts from Lucien Febvre's or Henri Pirenne's ideas. Does the methodology called *géographie humaine* come directly from Pierre Vidal de la Blache, who was not involved with *Annales* or through Lucien Febvre's book *La terre et l'évolution humaine* or perhaps through a Gernoble historian Raoul Blanchard, who lived in Quebec for many years? Dubuc (1978) does not undertake to answer these questions. These examples show that it is almost impossible to track indirect influences.

Another problem connected with the phenomenon of direct and indirect influence might be called "deferred influence": a methodological influence does not need to manifest itself right away – it might become visible after some time. The causes of this phenomenon might vary from institutional to intellectual ones. It should be noted that for example having taken

---

[3] Authorized testimony of Prof. Michał Tymochowski, 23 May 2005.
[4] Non-authorized testimony of Prof. Maria Bogucka, 11 June 2005.

cognizance with the subject matter of mentality and social beliefs during her first stay in Paris in the late 1960s, Maria Bogucka did not focus on those right away. It resulted partly from the institutional dependence on Prof. Małowist and from her willingness to make use of the surplus sources concerning the Baltic zone's economic history.[5] Michał Tymowski's testimony reveals further complications: the scholar states that now, in the early 21st century, he began working on the subject matter of the course of initial meetings between "savage" Africans and colonizers – how they saw each other, what gestures they made, etc.[6] This topic perfectly fits the trend of the *Annales* historical anthropology. Presuming that French achievements in that field did influence Polish Africanist's epistemology, we encounter the following question: when did it happen? For it seems unlikely that it took place at the beginning of the 21st century; perhaps the seed that is sprouting now was planted during his first visit to Paris in 1970.

Fallaciousness or, should we say, traps inherent in the analysis of mutual influences are visible in Leszek Kołakowski's short text *Historia jako sztuka piękna* (History as Fine Art) published in 1962. In his work, which is a loose review of Johan Huizinga *The Autumn of the Middle Ages*, the philosopher states that history is more like art than science. It differs from literature only by being based on sources and studies (Kołakowski 1989). These observations are strikingly similar to the ideas of Paul Veyne (1971), who about a dozen years later published a book that became a bestseller in France, in which he proposed almost identical theses. Thus, we might suspect that this time, quite unusually, the Frenchman was inspired by the Pole's work. It does not seem true, however, for a very simple reason: Veyne was surely unfamiliar with Kołakowski's marginal, few-pages-long text published on a larger scale only in 1989 in a collection of the philosopher's selected texts and, making matters worse, in Polish. Both scholars arrived at similar conclusions "mimetically" because they were at similar stages in their intellectual evolutions and had read similar books even though they knew little (or nothing?) of each other. Moreover, Leszek Kołakowski referred to the views of Benedetto Croce or Wilhelm Dilthey, who already a few dozen years earlier had also seen history as part of literature. Thus, he could have simply been inspired by the history of historiosophy, and not even the French one. On the other hand, as Jerzy Topolski points out, Paul Veyne himself was also influenced by Benedetto Croce, Robin George Collingwood or Henri-

---

[5] Non-authorized testimony of Prof. Maria Bogucka, 11 June 2005.
[6] Authorized testimony of Prof. Michał Tymochowski, 23 May 2005.

Irenée Marrou's anti-positivistic trend of "comprehensive history" (Drozdowicz, Topolski and Wrzosek 1990, p. 16). This would prove Veyne and Kołakowski's mutual inspirations.

If this evident similarity does not prove Kołakowski's influence on Veyne then it certainly testifies to the supra-provincial nature of the former's thoughts. Employing Paul Ricoeur's terminology again, it might be said that the two scholars belonged to one "hermeneutic circle" or that they shared "intersubjective understanding," to use Jürgen Habermas and Charles Taylor's terminology (see Taylor 1980; see also Drozdowicz, Topolski and Wrzosek 1990, p. 26.). It would also be valid here to mention the term used by Karl Mannheim (and by Andrzej Walicki in Poland), who wrote about "spatial-temporal locus" (see Sitek 2000, p. 158). Kołakowski and Veyne functioned in a similar locus in the intellectual time-space, which transgressed national barriers. By means of a comparative analysis of Polish and French historiography from a hermeneutic and epistemological perspective during communism one could come to the conclusion that despite the Iron Curtain's existence European culture (at least the scientific one) is universal. It is a more certain and perhaps more momentous conclusion than an affirmation of *Annales'* methodological influence in Poland.

It should be noted that the very fact of becoming (usually inadvertently) a member of the European guild of scholars, and consequently becoming a part of a certain international, supra-provincial "subculture" of historians – achieved by reading, contacts and travelling – happens in a context at least partially separated from the cultural context of the given scholar's background. The possibility of "making oneself understood" and the intellectual-methodological similarities stem from a certain cultural compromise, from liberation from the subjectivity of one's background. This compromise and the cultural abstraction themselves imply further uniformization of opinions and the ways of their articulation (cf. Bourdieu 1984, esp. pp. 99-167). I call this phenomenon the "uniformization of poetics." Even though it cannot be absolutized, it does seem, however, that it is due to this that we can explain the similarities in Polish and French historians' works more precisely and clearly than we could using the theory of influences. If the above sentence is too far-fetched a statement then let us at least state that without uniformization and universalization there are no influences: in order to influence one another, scholars need to make themselves understood. Mutual understanding requires use of a similar terminology, or more broadly speaking: use of a similar cultural code. And a similar cultural code might produce similarities in the scientific code.

As I have mentioned, some scholars of this phenomenon yield to the influences' fallaciousness. It most often involves the identification of a given work discussing issues similar to the Annalists' work, with an indeterminate "*Annales* spirit." Yet not all works of that character have to have that "spirit." The huge intellectual impact of the *Annales* milieu, combined with exceptionally solid institutional foundations in the form of VI EPHE/EHESS Section, was conducive to reasoning in a category of a certain "annalistic holism." Additionally, scholars were inclined to think that everything that was similar to the Annalists' achievements had to stem from their influence, and this is a very dangerous trap which must be avoided. Making matters worse, many scholars, especially in the 1950s, 1960s and 1970s declared that their work had that "spirit." That was part of the game they played to increase their "symbolic capital" (to use Pierre Bourdieu's term) – presenting oneself as a scholar closely connected with the famous group increased one's prestige in the milieu (see e.g. "Discussion," in Birnbaum 1978).

Does the existence of the Western European intellectual community and of the "scientific International" offer a better explanation of the similarities between Polish and Annalistic historiographies than the one offered by the thesis about the influence the latter historiography had upon the former?

Italian scholar Mario del Treppo's (1977) highly crucial statement voiced in the mid-1970s, quoted by Maurice Aymard, partly answers that question. And so in Italy there was

> a common historiographic practice – inspired by criteria of globality and interdisciplinarity – of a considerable degree of research prospects' openness, parallel to the great variety and specialization of research techniques; an eclectic and syncretic historiography, which could adopt Jerzy Topolski's *Methodology of History* (first published in 1966, then in 1973, Italian translation in 1975) as its central work, as a certain scholastic *summa*. (quoted after Aymard 1978, p. 59)

According to this highly interesting opinion, Polish historians were part of that historiographic trend on equal rights but actually its role was paramount. Hence, Mario del Treppo did not regard the role of *Annales* as dominant. At the same time it remains obvious that most European researchers did not read Jerzy Topolski's work. Let us notice that an Italian translation was published only in 1975, when the model of historiography described by del Treppo was just being superseded by new trends. Hence, it seems that in this statement the Polish scholar's work is treated as a certain symbol of the intellectual community in which Poles and Annalists played a significant role.

How could we define that community more precisely? Using the science model created by Thomas Kuhn, in his monograph devoted to the Anglo-Saxon new economic history, Jan Pomorski listed the points he used while recreating that historiographic school's paradigm. Firstly, he had to isolate the "scientific community" making up that trend or milieu. That was done with the use of the following sociological tools: the scholar tracked down similar education of a community, a similar course of scientific initiation, inspirations, affiliation with the same associations, periodicals, participation in the same conferences, etc. Having defined the "scientific community," he went on to specifying the community's common paradigm – i.e. a vision of the world ("compositional metaphysics"), as well as shared "symbolic generalizations" (a specific way of expressing shared laws of science), common values and models of solving problems (Pomorski 1995, pp. 24-29; see also Radomski 1999, pp. 130-131).

Using Thomas Kuhn and Jan Pomorski's methodology, we might say that in the case of (some) Polish and French non-classical historians it would be justified, with many reservations, to talk about the formation of one, supra-provincial "scientific community" not limited solely to the representatives of these two countries. A certain number of historians from Poland and France (the upper limit being the total number of the visiting Polish researchers) knew one another in person, participated in the same scientific events, belonged to international scientific associations, shared some historiosophic-epistemologic and technical views. Of course, diametric differences in political and ideological realities as well as in every-day life in Poland and France also played their role. Consequently, referring to the above-mentioned scholars' terminology, it is best to talk about a "loose scientific community." The too far-fetched thesis about a common paradigm of the historians of École and of the Tadeusz Manteuffel Institute of History of the Polish Academy of Sciences cannot be upheld due to these differences.

It is precisely this "loose scientific community," operating thanks to partial uniformization of poetics, which implies a transgression of provincialisms and similar methodological choices. As I have suggested, it might stem not so much from influence of one group on the other as from the phenomenon of "mimesis." "Methodological mimesis," understood as independent arrival at similar historiographic conclusions, most often results from an existence of an intellectual community – from "the spirit of the times." It is this *Zeitgeist* that is the foundation of "mimesis," if not the condition. It cannot be ruled out that two scholars from two totally different intellectual families can indeed arrive at similar ideas, but in the case of Polish and French humanists it is sounder to mark the role of "the spirit of the times." Only in terms of intellectual affinity can we most fully explain why for example Krzysztof Pomian and Jacques Le Goff had independently taken up pioneering research in the field the of perception of time before they became familiar with each other's work (see Pomian 1998, esp. pp. 76-79). We have evaluated the similarities between Leszek Kołakowski's and Paul Veyne's ideas in the same manner. There are many examples of such "pairs."

We should discuss two other characteristic features of "methodological mimesis" based on the example of Witold Kula's work. The first one involves Polish and French historians' use of similar wording, expressions, terminology. And so Witold Kula in his book *The Problems and Methods of Economic History* (Kula 2002), uses terms such as 'economic situation' (e.g., pp. 203-204) or 'economic cycles' (e.g., pp. 733-734). Fernand Braudel (1996) used similar expressions while writing for example about the "average" speed of historic time in the second volume of *The Mediterranean*. We might wonder whether the common use of the similar terms resulted from mutual influences or simply from reading the same books on economy. The latter is surely easier to prove. But the situation becomes clear if we assume the existence of the phenomenon of uniformization of poetics.

The other phenomenon, equally characteristic of "methodological mimesis," is connected with the use of similar wording. I call it the "internalization of quotations." What I mean here is the habit of interspersing one's reasoning with the quotation of other scholars' works. Writing about the historian's cognitive capabilities in his *Rozważania o historii* (Reflections on History), Witold Kula refers to the *Annales*' circle. "*Toute histoire est choix*," as Lucien Febvre has said. It is deeply true. "All history is choice" (Kula 1958, p. 67). The Polish historian "internalized" his French colleague's view: right after the quotation he repeats the Frenchman's thought without the quotation marks. Is it an evidence of an influence? Surely this measure proves that Witold Kula agreed with Lucien Febvre in that regard but both scholars could have independently arrived at the same conclusions. Such an admission of similarity in opinions would be a classic example of "inspiration."

It seems that we should look in a similar vein at Kula's reflections on geography as a subject of interest for the historian. He stated that "changeability of natural phenomena must not be treated as extra-historical (extra-social) processes." After these words he referred to Braudel's *The Mediterranean* (Kula 1978, p. 89; see Piasek 2004, pp. 60-61). It could once again be evidence of a mere similarity in opinions. Kula, in turn, wrote e.g. "I completely agree with what was said . . . by Paul Ricoeur. Historian's work is a permanent effort to translate the values

created by man from one language to another" (Kula 1988, p. 147). So did
the scholar find in Ricoeur just a confirmation of his views or did he take
them over and began considering them his own when he wrote later that he
"agreed" with them?

Of course, the "internalization of quotations" is not characteristic
solely of Witold Kula. Pondering over the complicated issue of the
historian's epistemology and subjectivism, Andrzej F. Grabski writes:

> the historian commences work on this or that topic because he thinks that it
> is important for some reason, hence, to some extent, he refers to some
> valuation at the very beginning of his work. "Each piece of historical
> research," wrote the great French historian Marc Bloch, "postulates that it
> needs to proceed in a specified direction from the very first step . . ."
> Asking a question is of fundamental importance in scientific cognition; this
> question's significance is nowadays revalorized by modern epistemology
> which accentuates the active role of the cognizing subject in the cognitive
> process. (Grabski 1985, p. 25)

It was necessary to quote the whole excerpt to show how the quotation
from Marc Bloch became an integral part of the Polish scholar's
exposition. However it remains unanswered if Grabski's line of reasoning
would have been the same without Bloch.

Certain manifestations of the "internalization of quotations" might be
explained in an easier way. Andrzej F. Grabski wrote in his article:

> we all remember Jerzy Topolski's essayistic account squaring with
> contemporary attempts at creating a "world without history," hence I shall
> repeat a thought not after him, but after P[ierre] Chaunu, which many of us
> would perhaps regard as their own: "A developing society must pay
> attention to its history; for history is a *condicio sine qua non* of setting the
> progress in motion and making people aware of it; it provides milestones of
> time measurement, which allows one to see the way he has come."
> (Grabski 1981, p. 34)

It is easy to notice that Jerzy Topolski wrote his book in 1972, while
Pierre Chaunu wrote his three years later. Thus, the quotation from
Chaunu was to show that similar opinions – about the danger of regarding
history as useless – are shared in various countries. Hence, the quotation
and the accompanying declaration were not a admission to yielding to the
influence of Chaunu's thoughts, but it was rather to convey a message that
"elsewhere people think similarly to us." Perhaps its task was to
accumulate the "symbolic capital" – emphasizing that Annalist's views
had a dignifying role.

* * *

The reflections on "intellectual mimesis" suggest that Pierre Bourdieu's opinions presented in the beginning are perhaps too pessimistic and that historical sciences do not need to exist solely within a provincial framework. In so far as the major evidence of universalism: methodological influence can, as we have seen, be easily rebutted, then the methodological similarities might as well testify to "intellectual affinity" between a certain group of Polish historians and the *Annales* milieu. We should, however, bear in mind that this "certain group" constituted just a small percentage of the total number of Polish students of history. Even such a unique and rich offer of a scholarship program as Braudel's engaged just a small percentage of scholars from the People's Republic of Poland. What about the rest? Surely we cannot automatically rank them among provincial scholars but their misunderstanding of the *Annales* methodology is quite certain. Let the quotation from Witold Kula's diary be a starting point to a discussion on that subject. This is how he reported on Fernand Braudel's lecture at the Warsaw University Institute of History on 9 November 1962:

> if only it could have been a lecture exclusively for Olek [Gieysztor], Geremek, me and friends!
> Unfortunately, in the room there were mostly mammoths, ichthyosaurs, rhinos, etc. All those for whom only *Jogaila's Hodoeporicon* constitutes serious science. (Kula 1996, p. 124)

Institute of National Remembrance
Public Education Office, Warsaw Branch
ul. Stawki 2
00–193 Warszawa
Poland
e-mail: patryk.pleskot@ipn.gov.pl

REFERENCES

Aymard, M. (1978). The Impact of the Annales School in Mediterranean Countries. *Review* **3-4**, 59.

Birnbaum, N. (1978). The Annales School and Social Theory. *Review* **3-4**, 236-242.

Bourdieu, P. (1984). *Homo academicus*. Paris: Les Éditions de Minuit.

Bourdieu, P. (2002). *Les conditions sociales de la circulation internationale des idées*. *Actes de la recherche en sciences sociales* (December 2002), 3-8.

154   *Patryk Pleskot*

Braudel, F. (1996). *The Mediterranean and the Mediterranean World in the Age of Philip II*. Vols. 1-2. Berkeley and Los Angeles: University of California Press.

Burke, P. (1978). Reflexions on the Historical Revolution in France: The "Annales" School and British Social History. *Review* **3-4**, 150-151.

del Treppo, M. (1977). Introduction. In: M. Cedronio, F. Diaz and C. Russo (eds.), *Storiografia francese di ieri e di ogii*, p. xxiii. Napoli: Guida.

Drozdowicz, Z., J. Topolski, W. Wrzosek (1990). *Swoistości poznania historycznego*. Poznań: Wydawnictwa Naukowe UAM.

Dubuc, A. (1978). The Influence of the *Annales* School in Quebec. *Review* **3-4**, 124.

Gostkowski, Z. (1965). *Le programme d'études à Paris*. Centre des Archives Contemporaines, Dossiers des Polonais, 1988 0326, art. 25.

Grabski, A.F. (1981). Modele historiografii współczesnej. *Przegląd Humanistyczny* **4**, 233-249.

Grabski, A.F. (1985). Historia a nowoczesność. In: *Kształty historii*. Łódź: Wydawnictwo Łódzkie.

Kołakowski, L. (1989). *Historia jako sztuka piękna*. In: *Pochwała niekonsekwencji*, vol. 3, pp. 152-155. London: Puls.

Kula, W. (1958). *Rozważania o historii*. Warszawa: PWN.

Kula, W. (1988). Obiektywizm badania historycznego a wartościowanie. In: *Wokół historii*, pp. 146-148. Warszawa: PWN.

Kula, W. (1996). *Rozdziałki*. Ed. M. Kula. Warsaw: Trio.

Kula, W. (2002). *The Problems and Methods of Economic History*. Aldershot: Ashgate Pub. Ltd.

Piasek, W. (2004). *Antropologizowanie historii. Studium metodologiczne twórczości Witolda Kuli*. Poznań: Wydawnictwa Poznańskie.

Pleskot, P. (2010). *Intelektualni sąsiedzi. Kontakty historyków polskich ze środowiskiem "Annales" 1945–1989*. Warszawa: Wydawnictwa IPN.

Pomian, K. (1978). The Impact of the Annales School in Eastern Europe, *Review* **3-4**, 101-121.

Pomian, K. (1998). Temps, espace, objets. In: J. Revel and J.C. Schmitt (eds.), *L'ogre historien: autour de Jacques Le Goff*, pp. 73-84. Paris: Gallimard.

Pomorski, J. (1995). *Paradygmat "New Economic History." Studium z teorii rozwoju nauki*. 2nd edition. Lublin: Wydawnictwo UMCS.

Radomski, A. (1999). *Kultura – Tekst – Historiografia*. Lublin: Wydawnictwo UMCS.

Sitek, R. (2000). *Warszawska szkoła historyków idei. Między historią a teraźniejszością*. Warsaw: Scholar.

Taylor, Ch. (1980). Understanding in Human Sciences. *Review of Metaphysics* **34**.

Veyne, P. (1971). *Comment on écrit l'histoire*. Paris: Seuil.

Wierzbicki, A. (2004). W stronę postaw otwartych. "Annales" w refleksji historyków polskich po II wojnie światowej. *Klio Polska. Studia i Materiały z Dziejów Historiografii Polskiej po II Wojnie Światowej* **1**, 49-81.

Wenceslao J. Gonzalez

METHODOLOGICAL UNIVERSALISM IN SCIENCE AND ITS LIMITS

IMPERIALISM VERSUS COMPLEXITY[*]

ABSTRACT. Universalism in science, when conceived in methodological terms, leads to the problem of the limits of science. On the one hand, there is "methodological imperialism," which in principle involves a form of universalism. On the other hand, there is the multivariate complexity – structural and dynamic, as well as epistemological and ontological – which represents a huge problem for methodological universalism, as may be seen with the obstacles for scientific prediction. Within the context of the limits of science, there is a better understanding of the issues of expansionism and imperialism.

One influential tendency in science is methodological universalism, which involves the attempt – explicit or implicit – to reach a whole realm of scientific research.[1] In some cases, it might be understood as

---

[*] This research is supported by the Spanish Ministry of Science and Innovation (FFI2008-05948). The research project deals with the sciences of complexity, which are particularly interesting from the point of view of limits of science. In this regard, some aspects related to topics of the present paper can be seen in Gonzalez (2011a).
[1] There might be interdependence between universalism in methodology of science and other kinds of universalism, mainly in epistemology and ontology of science. Thus, universalism regarding the processes of scientific research might be accompanied by

In: Krzysztof Brzechczyn and Katarzyna Paprzycka (eds.), *Thinking about Provincialism in Thinking* (*Poznań Studies in the Philosophy of the Sciences and the Humanities*, vol. 100), pp. 155-175. Amsterdam/New York, NY: Rodopi, 2012.

"methodological imperialism," an approach that includes a kind of dominion on the processes to be used about certain aspects of reality. This view broaches the problem of the limits of science, which is connected to the difficulties posed by complexity in its different forms: structural and dynamic, epistemological and ontological . . . Thus, within this context of limits of science, there might be a reflection on expansionism and its differences with methodological imperialism. These aspects require following several steps in the philosophical analysis.

Firstly, there is a consideration of the varieties of methodological universalism, which gives us a certain framework for "methodological imperialism." This step involves the existence of several levels of methodological analysis and takes into account the problem of the historical dimension of science. Secondly, there is a reflection on the limits to methodological universalism due to the problem of complexity. This step leads to the existence of obstacles for this approach and, additionally, for methodological imperialism. Moreover, these obstacles to methodological universalism are particularly relevant in the case of scientific prediction. Thirdly, we can think of how the limits of science might shed light on expansionism and imperialism in scientific research.

1. Varieties of Methodological Universalism

Prima facie, there are several possibilities of methodological universalism in science,[2] which are in tune with the scope of the research. Thus, these methodological attempts can be developed at several levels and from different angles, according to the aims of the research, the kind of processes to be used, and the expected results of this human activity. In the most general cases, they can be seen as an expression of the search for unity in science, which is a perspective that appears from time to time in the characterizations of the methodology of scientific research.[3] But, when

---

universalism in epistemological claims (empiricist, rationalist, realist, . . .) and in the characterization of the reality that is being researched.

[2] "Universalism" in the methodology of science has some similarities with the concept of "perfect science" in epistemology, insofar as they can be seen as an "ideal" rather than as a "goal." The criticism of the idea of a "perfect science" can be found in Rescher (1999), chapters 4, 5 and 6, pp. 98-148; and Rescher ([1984] 1999). On his views, see Almeder (2008).

[3] Unity and diversity of science are usually present in the methodological analyses. They are at stake in the well-known methodological comparison between the natural sciences and the social sciences, cf. Gonzalez (2003a). Frequently, in the comparison between both groups of

the scope is less general, methodological universalism can be just the way for a scientific enlargement, in order to solve new problems or to deal with novel facts (i.e. new facets of reality).

Within the different levels of scope of scientific research, these proposals of methodological universalism are sometimes *explicit* (e.g. in the ideal of the unified science), whereas at other times the approach is rather *implicit* (e.g. in the case of certain evolutionary perspectives, whose views are sometimes broadened beyond initial expectations); see, for example, Hofbauer and Sigmund (1988) as well as Heylighen, Bollen and Riegler (2011). When universalism is explicit, there is the clear-cut acceptance of a method as valid, in principle, for a complete domain or, at least, for a large extension of it. Meanwhile, there are universal approaches that are such in a rather implicit way. This happens when there is an assumption of a dominant method (a "standard"), which might be accompanied *de facto* by a clear disregard for any other alternative method for a concrete domain.[4]

Both options of methodological universalism – explicit and implicit – can lead to a kind of "methodological imperialism,"[5] insofar as there is – in one way or another – an extremely influential or "official" conception of how to develop scientific research at the level discussed (i.e. science in general, a group of sciences, or a specific science). The intensity of the methodological imperialism increases if the methods are used in a territory that was previously considered within another realm. However, it might also be the case that a methodological imperialism is not universal, insofar as it is accepted that there are some phenomena that require a particular method, due to their concrete features (e.g. in evolutionary approaches to methodology of science).[6]

---

sciences, there is an insistence on the need for improving the social sciences, cf. Taagepera (2008).

[4] Among the economic schools, there are views that are considered as antagonists and, therefore, they are rejected by another school (as is the case between neoclassical methodological individualism and Marxist social holism). But there are also competing schools that have methodological differences that are not *eo ipso* incompatible regarding some phenomena (e.g. between Herbert Simon's behavioral economics and Oliver Williamson's new institutional economics). Cf. Simon (2001); Williamson (1975).

[5] See, for example, Dinga (2009). This option is also considered in other social sciences, cf. Kuorikoski and Lehtinen (2010).

[6] Natural selection was conceived by Charles Darwin as the key mechanism – the main cause or agency – responsible for all divergence on earth but not the only one (cf. Gonzalez 2008a, p. 12; Hodge and Radick 2003). Thus, if we assume this ontological element as a condition for a methodological approach to scientific research, then the evolutionary methodology of "selection of theories" can be non-universal.

1.1. *Levels of Methodological Analysis*

Regarding the varieties of methodological universalism and the possibility of a methodological imperialism, a key element is the extent of the scope, which allows several levels of methodological analysis. In principle, there are three *main levels* of scientific research: first, science in general (mainly, the empirical sciences)[7]; second, a group of sciences, such as the natural sciences, the social sciences, or the sciences of the artificial[8]; and third, specific sciences, such as biology, economics, or computer sciences. These levels can embrace the diverse options of a methodological universalism in scientific research.

These forms of methodological universalism might be considered as *methodological imperialism* in at least two different ways: intensity within a realm and extension to other realms. Thus, a) methodological imperialism can conceived in terms of a kind of neat prevalence or clear dominion of some methods regarding a certain scientific realm (such as happened historically with Newtonian mechanics within physics as a whole); and b) methodological imperialism can be understood as a set of methods that come from a different discipline, whose "boundaries" overflow to impinge on another field or fields (e.g. economic methods used in sociology, psychology, anthropology, law, political science, archeology, etc.). This predominance could be the case in any of these three levels of methodological analysis.

Certainly, methodological imperialism can be thought of as *science* in general, for example, developing a methodological proposal based on logical grounds and assuming the idea of universal validity of logic. But it is also possible to think of a methodological imperialism of a naturalist kind, based, for example, on evolutionary grounds, such as the influential Darwinian approach (cf. Gonzalez 2008a). Its repercussion is very noticeable in natural sciences, such as biology; in social sciences, such as in the evolutionary conceptions of psychology and economics (cf. Nelson and Winter 1982); and in the sciences of the artificial, such as computer sciences in terms of evolution of a complex system (cf. Simon [1969] 1996, pp. 188-190).

But, as a matter of fact, the attempt at a "methodological imperialism" has been made explicitly in the social sciences, while using economic methodology for solving very relevant social problems. This proposal for

---

[7] Formal sciences, such as mathematics, commonly have specific methodological considerations, even though "quasi-empiricist" approaches and naturalist conceptions have searched for methodological similarities with empirical sciences.

[8] These sciences are understood here in terms of Simon ([1969] 1996).

phenomena of a *group of sciences* has usually been connected to the work of a Nobel Prize winner in economics, Gary Becker,[9] a central figure of the Chicago school. Nonetheless, from a historical point of view, there are other authors that have been considered as supporters of an imperialism of economic roots.[10] Becker has tried to solve important social problems (e.g. those regarding family matters, such as marriages, divorces and fertility) by means of economic methods (based on neoclassical models).[11]

Besides the methodological universalism in a group of sciences, as has happened in the social sciences, this methodological proposal can also be located within the *specific sciences*. Thus, we can see this phenomenon in disciplines of the natural sciences, such as biology (which is largely influenced by processes understood in evolutionary terms, frequently Darwinian); in subjects of the social sciences, such as economics (where the methods of mainstream economics – sometimes called "orthodox" economics – are still extremely influential); and in studies of the sciences of the artificial, particularly in sciences of design developed in the sphere of artificial intelligence.

1.2. *The Historical Dimension*

Concerning these three levels of *methodological universalism* indicated (science, group of sciences, and specific sciences), there is a historical dimension to be considered in them. In this regard, the methodological contents can be seen in different ways, where two seem to be most relevant here: (i) a methodological universalism conceived in a somehow "timeless" format, where the universality is accepted across times (i.e. it can be enlarged but not revised in any strong sense), and (ii) a methodological universalism within the context of historicity, where there

---

[9] His view is analyzed in Pies and Leschke (1998). On methodological imperialism from a Popperian perspective, see Radnitzky and Bernholz (1987).

[10] This is the perspective defended by Uskali Mäki, who sees the imperialistic inclination of economics as an operation for the last half century. As examples of this trend, he cites "works such as Gary Becker's *The Economics of Discrimination* (1957), Anthony Downs's *An Economic Theory of Democracy* (1957), James Buchanan and Gordon Tullock's *The Calculus of Consent* (1962), Mancur Olson's *The Logic of Collective Action* (1965), and Becker's *A Treatise on the Family* (1981). As their titles indicate, these works deal with domains of phenomena that previously were not generally perceived as 'economic' but are now analyzed in economic terms" (Mäki 2009, p. 352).

[11] Among his most influential works are Becker (1976) and Becker (1981). On his views, see Cabrillo (1996). Regarding this topic, cf. Stigler (1984).

is a temporal universality and a variation of contents according to the advancement of science.[12]

A methodological universalism in a somehow "timeless" format – something that generally avoids the historical dimension of the methodology of science – might be proposed for any of the three levels of analysis: science, in general (focused on empirical disciplines), a group of sciences (mainly, natural sciences), or some specific sciences (especially, physics). This is what can happen in the case of a methodological conception directly based on a logical support, which was the program developed by logical positivists (mainly in the early stages of this intellectual movement during the Vienna Circle),[13] which was softened by some logical empiricists (cf. Parrini, Salmon and Salmon 2003; Rescher 2006). In this regard, the ideal of the "unified science" can be seen as a methodological universalism rooted in logical grounds, where physics appears as a key science (i.e. as an archetype or exemplar for other sciences) within the sphere of the empirical sciences.[14]

Another version of a supposed methodological universalism is the acceptance of the "scientific method," a label that was part of Karl Popper's professorship at the London School of Economics ("Logic and Scientific Method"). Moreover, some still believe that Popperian falsificationism was a sort of universal methodological approach. But, firstly, Popper himself did not assume this idea of "the" scientific method, understood as a systematic way to achieve scientific results well founded (cf. Worrall 2001, p. 114). Secondly, he did not propose *de facto* a "Logic of scientific discovery" (i.e. a methodology of logical roots for making discoveries).[15] Thirdly, his general rules for methodology were oriented toward the questions of validation of scientific discoveries (Popper [1935] 1959, p. 31), which in his falsificationism is made by means of the critical elimination of error. Fourthly, Popper proposed the method of "situational

---

[12] On the role of *historicity* in science and the differences with "evolution" and "revolution" regarding the characterization of the scientific change, cf. Gonzalez (2011b).

[13] On the logical positivist program, see Carnap (1938-1955) and Ayer (1959, 1981); see also Friedman (1999). The paradox is that a key alternative to this program oriented toward stability in science, due to the logical foundations, was published within the series of the encyclopedia devoted to unified science: Kuhn ([1962] 1970).

[14] See, for example Carnap (1932). The attempt at a unified conception can follow other paths as well, such as cognitive grounds, cf. Giere (1984).

[15] "The initial stage, the act of conceiving or inventing a theory, seems to me neither to call for logical analysis nor to be susceptible of it" (Popper [1935] 1959, p. 31).

analysis" and the principle of rationality for the social sciences, which involve important methodological differences with natural sciences.[16]

Nevertheless, there is the possibility of a methodological universalism that can be expressed in historical terms. This might be the case if influential scientific communities assume methodological universalism *de facto* during some periods. This can happen in the history of science, when some methods become dominant in a specific science or group of sciences for a number of years. This interpretation is possible, for example, in the case in the past regarding methods used in mechanics, which were generalized for other parts of physics and even for other disciplines (on this topic, see Agazzi [1969] 1974), including social sciences such as sociology and economics.[17]

What is more, it seems clear that the forms of methodological universalism in science can change historically. Hence, a dominant methodological approach in a particular subject can be substituted by a subsequent methodological perspective, which is seen as more fruitful within a specific domain and became all-prevailing in the field for some years or a few decades. Thus, the methodological universalism of an "imperialist" kind can be temporal, because another conception can be developed that eventually replaces the old dominant methodological perspective. Therefore, imperialism might be "universal" just for a while, but it hardly can be an approach that is somehow "historically permanent." The perennial option seems almost impossible due to the *critical attitude* in science (cf. Niiniluoto 1984), which involves constant revision of the processes in order to achieve new aims.

According to this historical variation regarding methodological universalism, it may be the case in many disciplines, such as economics, that there is a transition from extremely influential methodological approaches to new conceptions, which are relevantly different. An example is the overwhelming dominance of economic models based on the rationality of the agents understood as maximization of subjective expected utilities (a key approach of the neoclassical view). Step by step, these models are increasingly replaced by an alternative view – proposed by Herbert Simon – which is better grounded from the empirical point of

---

[16] This can be seen in an analysis of his views on scientific prediction, cf. Gonzalez (2004).
[17] There is a well-known attempt by August Comte to build up sociology as "social physics." In addition, the influence of classical mechanics on classic economics is commonly assumed. Cf. Mirowski (1989).

view, and where the economic methods are built on rationality of the economic agents conceived as bounded.[18]

2. The Limits for Methodological Universalism:
The Problem of Complexity

The whole variety of expressions of methodological universalism needs to deal with the problem of *complexity*. This facet involves the issue of limits of science and it is also crucial for dismissing systematic attempts at elaborating a "methodological imperialism." It seems clear that there are structural forms of complexity and dynamic aspects of complexity that, in principle, are beyond the proposals of a dominant set of methods in a group of sciences or within a specific science.[19] In effect, within the structural complexity and the dynamic complexity we can see epistemological and ontological aspects of complexity.[20] They certainly have a direct repercussion on the processes of scientific research and its possible limits.

Moreover, the amount of difficulties posed by both kinds of complexity – structural and dynamic, with their epistemological and ontological elements – have called frequently for enriched methodological approaches, such as those developed in terms of multidisciplinarity and interdisciplinarity. Thus, we can think of the contributions of several disciplines as if they were "layers" on a topic of research (a multidisciplinarity viewpoint) or we can consider the convergence of the contributions of a set of disciplines toward a "meeting point" of their scientific research (an interdisciplinarity perspective).

2.1. *Obstacles to Methodological Universalism Due to Complexity*

Complexity is indeed a very important source of difficulties for methodological universalism because it affects problems, methods and results of the scientific research. The features of complexity can be

---

[18] There is now an important list of Nobel laureates in economics in favor of "bounded rationality": Herbert A. Simon (1978), Douglass C. North (1993), Reinhard Selten (1994), Daniel Kahneman (2002), Robert J. Aumann (2005), . . . New aspects of this topic can be seen in Gonzalez (2008b).
[19] On the structural complexity and the dynamic complexity, see Gonzalez (2011a, pp. 321-325).
[20] Regarding epistemological and ontological forms of complexity, cf. Rescher (1998a, pp. 1-26; especially, pp. 8-16).

thought of as science, in general, a group of sciences, or a specific science because complexity can be considered, in principle, by any of the disciplines related to nature, social and artificial worlds (see, in this regard, Mainzer ([1994] 2007). To some extent, we can consider complex systems in these main spheres of reality.

Furthermore, complexity can be focused either from the structural perspective or from the dynamic viewpoint,[21] which involves the possibility of emergent properties.[22] In the first case, the study of complexity is commonly made regarding the framework or constitutive elements present in a group of sciences or in a specific science, whereas in the second possibility, the analysis of complexity is related to change over time of the motley elements involved in that collection of sciences or the specific science, taking into account the forces generating the change.[23]

Obstacles to methodological universalism can be located on both sides: in structural complexity and in dynamic complexity. Concerning the first case, we can take into account the main epistemological and ontological aspects. Nicholas Rescher (1998a, p. 9) has made a relevant presentation, where the *epistemic modes* of complexity are divided in three groups, where it is possible to find a formulaic complexity: 1) descriptive complexity, 2) generative complexity, and 3) computational complexity.

Ontological modes of complexity are distributed in three main groups: a) compositional complexity, b) structural complexity (in a strict sense), and c) functional complexity. Within the compositional complexity, the possibilities are twofold: constitutional and taxonomical (or hetero-geneity). Meanwhile, for Rescher, "structural complexity" also has two possibilities: it includes complexities organizational and hierarchical; whereas functional complexity is articulated in two modes: operational and nomic.

Rescher's analysis is mainly related to structural complexity (the complex framework of the elements of science) rather than focusing on dynamic complexity (that connected to scientific change). Nevertheless, he is open to some dynamic aspects, which are relevant for science, in general, a group of sciences or a specific science. These dynamic aspects might be detected in the generative complexity (in the epistemic modes of

---

[21] On complexity from a *dynamic point of view*, see Gonzalez (forthcoming).

[22] "The prospects for the emergence of an effective complex system are much greater if it has a nearly-decomposable architecture" (Simon 2001, p. 82).

[23] These categories of structural and dynamic can be used to articulate lists of kinds of complexity such as "multilevel organization, multicomponent causal interactions, plasticity in relation to context variation, and evolved contingency" (Mitchell 2009, p. 21).

complexity) and in the operational complexity as well as in nomic complexity (in the ontological modes of complexity).

Each mode of complexity – epistemic or ontological – can pose some difficulty for the universal methodology, regarding generality and reliability. The methodology of science needs to deal with issues that are not simple, which might be at different epistemological levels and can belong to diverse stages of reality. The researcher uses processes that depend on the objects (the aspect of reality) and the kind of problem (the focus of attention). Insofar as the scope is larger, the validity of the contents can, in principle, decrease due to the problems of testing the hypotheses.

Thus, if we think of a science such as economics, these sources of complexity resemble a scale with several steps: (i) the social and artificial realms; (ii) the micro and macro levels; (iii) the degree of autonomy as human undertaking ("economic activity" and "economics as activity"); (iv) the organizations and markets; (v) the role of individual agents (i.e. creativity in different realms); etc. Besides the structural complexity is the dynamic complexity, where historicity has a key role and is another obstacle for methodological universalism. The change introduced by historicity – in knowledge and in reality – makes it more difficult to get universality across historical periods.

2.2. *Methodological Universalism and the Obstacles to Predictors from the Angle of Complexity*

Methodological universalism needs to consider that basic science and applied science does not work in the same way, insofar as they have different aims, processes and results (cf. Niiniluoto 1993; 1995). Basic science commonly explains and predicts, while applied science requires predictions as a previous step for making prescriptions (cf. Simon 1990; Gonzalez 1998). Thus, both kinds of science share the importance of prediction: in the former case, prediction is crucial for increasing our knowledge of the world (natural, social or artificial); whereas, in the latter, prediction gives us knowledge about a possible future, a content that can be used as a guide for practical action.

Undoubtedly, there are general obstacles for a universal methodological approach to scientific prediction, which are worthy to be considered. These obstacles are mainly epistemological and ontological, and they require taking into account the angle of complexity, both structural and dynamic. In this regard, the case of economics as specific science seems particularly relevant, insofar as it has been used for conceptions of "methodological imperialism." But economics poses problems concerning the limits of science, which are more noticeable in making predictions due to issues such as the reliability of the knowledge of future phenomena.

Clearly, the obstacles for methodological universalism come also from the complex methodological status of economics, because it is a dual science: artificial and social (cf. Gonzalez 2008b, pp. 165-186 and, especially, pp. 166-171). This dual character of economics can be recognized insofar as it is a science of the artificial that uses designs to enlarge human possibilities now and for the future; and, at the same time, economics is a social science that deals with human needs in a cultural milieu (food, housing, clothing, etc.).

Artificial and social are, then, two features that increase the degree of complexity of economics in comparison with other sciences. Thus, the structural complexity and the dynamic complexity of economics require dealing with the combination of human designs and social needs. Both have a relation to the possible future and, therefore, with making predictions. In this regard, it seems clear that the dynamic complexity of economics poses undeniable difficulties for the reliability and accuracy of economic predictions. Hence, the analysis of complexity in economics, understood as an influential factor of difficulties for economic predictions, leads to the problem of the obstacles for making scientific predictions in economics.

Indubitably, this issue is central for economics, especially if we accept that prediction is or might be a test for economics as a science.[24] This is an issue related mainly to economic theory (i.e. "positive economics"), but there is also the task of prediction concerning applied economics (i.e. "normative economics"). This involves the view that, within economics, prediction plays a key role as "basic science" (and, frequently, it has been the focus of the discussions of the scientific character of economics; cf. Gonzalez 2006a) and as an "applied science."

Concerning this thematic sphere, some points should be considered about prediction from the point of view of complexity and methodological universalism. (i) Prediction is not a simple concept, especially from a *methodological point of view*, because its use in economics is twofold: prediction is a test for evaluating theories and it is also a guide used for policy. Thus, prediction needs to address a diversity of problems related to the knowledge of the future, both as basic science and as applied science. (ii) Those problems lead to *obstacles* to scientific prediction. Moreover, they are difficulties for a methodological universalism even within

---

[24] An analysis of the main views on this issue is in Gonzalez (2006b).

economics. *De facto*, there are diverse kinds of limits on scientific prediction in social and artificial realms such as economics.

Indeed, prediction is not a simple concept insofar as it deals from the outset with a variety of methodological problems, mainly in the social sciences:

> The nature and complexity of what one extrapolates from, the precision with which the processes of development are thought to be known, whether the outcome predicted has a contaminating effect on the prediction in question and may thus modify it, how far into the future this extrapolation is intended to predict, the range of variables which can be accommodated in calculations: all these are some of the many and more obvious problems which make foretelling the future a hazardous business. (Howe 1993, p. 4)

Following the methodological difficulties for prediction in economics, we can reach the epistemological problems and the ontological basis. In this regard, there are limits to scientific predictions in economics. These could be for "internal" aspects related to scientific activity or for "external" features of human capacities or social capabilities. The former belong to the sphere of the *constituents of science* and the limits indicated by them (semantic, logical, epistemological, methodological, ontological, axiological and ethical), whereas the latter are in *human agents* and their institutions.

Consequently, within the framework of a social and artificial science, economic models need to deal with the nature of things where choice or chance – and even chaos – can have a role, which can lead to creativity in human undertakings (e.g. the creation of artificial products like "hedge funds"). At the ontological level, Herbert Simon has pointed out that we "don't know whether the economy is a chaotic system" (Simon 1989, p. 99). Meanwhile, at the epistemological level, our cognitive limitations are undeniable, as the bounded rationality approach has repeatedly emphasized.

From a general point of view – focusing mainly on ontological and epistemological aspects – Nicholas Rescher calls attention to the *principal impediments* to predictability:

1) anarchy, where there is lawlessness in the strict sense of the absence of lawful regularities to serve as connecting mechanisms;

2) volatility, when there is an absence of nomic stability and then of cognitively manageable laws;

3) uncertainty, which is the lack of information about the operative mechanisms[25];

4) haphazardness, when the lawful linking mechanisms do not permit the secure inference of particular conclusions: a) *chance* and *chaos* (stochastic or random processes which make laws at issue irretrievably probabilistic),[26] b) *arbitrary choice* (determinations that are basically groundless and so rationally intractable), and c) change and innovation (a kind of novelty that make outcomes not foreseeable because prediscernible patterns are continually broken);

5) fuzziness, which is data indetermination whether individually or in a collectively conjugate way;

6) myopia, which is data ignorance in the sense of lack of sufficient volume and detail to be able to make a prediction;

7) inferential incapacity, which is the infeasibility of carrying out the required reasoning (cf. Rescher 1998b, pp. 134-135).

Noticeably, these impediments to predictability have a relation with complexity, and they pose difficulties to a methodological universalism. Some of these impediments are mainly structural, whereas others are clearly dynamic. Rescher recognizes that, for many writers, "complexity is determined by the extent to which chance, randomness, and lack of lawful regularity in general is absent" (Rescher 1998a, p. 8). But this concept the inverse of simplicity – admits of *degrees*: the system can be more or less complex. In the case of economics, the tendency is to focus on some of the previous impediments to predictability, where uncertainty has a key role from the methodological point of view (mainly in applied economics).[27]

---

[25] We can distinguish between the uncertainty that we are aware of ("what we know that we do not know") and the uncertainty that is actually beyond us ("we do not know that we do not know").

[26] The topic of chaos has received an enormous attention for years and exceeds the limits of this paper. Regarding its relation to complexity besides the book of C.S. Bertuglia, and F. Vaio (2005), see Smith (1998); and Strevens (2003). Chaos "is a feature of certain dynamical models which exhibit sensitive dependence on initial conditions plus 'confinement' plus (typical) aperiodicity. Which is to say, roughly, that tiny differences in the initial states can exponentially inflate into big differences in later states, but the values of the relevant state variables eventually remain confined within the fixed boundaries although typically never exactly repeating" (Smith 1998, p. 20).

[27] In the case of prediction, the methodological role of uncertainty is very important. Thus, once outcomes of economic forecasts are known, "the corresponding forecasts errors and the anticipated forecast uncertainty can be used to evaluate the models from which the forecasts were generated" (Ericsson 2002, p. 19).

While economists point out central obstacles for an adequate or reliable prediction, there are some differences in emphasis according to the diverse economic schools. These issues are usually connected with central epistemological and methodological controversies: the possibility of economic laws, the way of understanding "causality" in economics, the role attributed to economic creativity of individual agents, etc. They are also connected to ontological aspects, such as the kind of novelty in the real world.

But insofar as these economists are aware of doing science, which is both social and artificial, the attitude of economists is commonly to emphasize some "stable" (epistemological and methodological) elements: human rationality in the decision-making (cf. Gonzalez 2003b), the capacity to gather and observe unobvious regularities (cf. Hoover 2002, p. 173), etc. Therefore, economists try to overcome these obstacles of predictability and, at the same time, they are well aware of the present stage of economic predictions (cf. Clements and Hendry 2002; see also Franses 2006).

What seems clear is that, ultimately, there are ontological roots: the difficulty of the methodological problem of prediction in economics – its necessity and unreliability – lies in the *complexity* of human activity in the social setting.[28] This complexity contributes to the frequent unreliability of economic predictions, which have their roots in the subject matter of this science: economic reality is a social and artificial undertaking, which is commonly mutable as a consequence of its dependence on human activity that is historically developed.

Nevertheless, "economic activity" is, in principle, an objective and measurable reality. In addition, economics as an *activity among others* is also objective but more difficult to measure.[29] Most of the econometric models of neoclassical economics focus on "economic activity," something that can be considered in itself (cf. Morgan 1990). But the ongoing economic crisis reveals the need to pay attention to the complexity of economics as an activity interconnected with other activities (social, political, cultural, ecological, etc.).[30] These aspects (ontological,

---

[28] "The economy consists of the activities of many millions of decision makers, acting largely independently but sharing information used in forming their decisions: the economy is thus very complicated" (Granger 2001, p. 93).

[29] On the distinction between "economic activity" and "economics as activity," see Gonzalez (1994), pp. 253-294; especially, pp. 261-280.

[30] Some economists, such as Joseph Stiglitz (Nobel Prize winner in 2001), insist on the need for new parameters for economics: the mere values of market activity are not good enough

epistemological, etc.) related to economic life make it difficult to have a methodological universalism, due to the number of factors at stake.

3. Coda: Limits of Science, Expansionism and Imperialism

If the problems of complexity in science pointed out here are taken seriously (which are especially relevant in the case of economics), then there are epistemological and ontological factors that work as limits for a methodological universalism. These limits are clear in two different directions: a) in the case of limits of science understood as "frontiers" or "barriers" (*Schranken*) regarding human knowledge;[31] and b) in the consideration of limits of scientific research from the perspective of a "ceiling" or a "final end" (*Grenzen*) of the search made by this human undertaking.[32]

On the one hand, a method works in a sphere of reality (even though it might be very large, as happens with physics in the case of relativity theory) and according to problems that might be tractable by the researchers (i.e. nobody does research "in general").[33] This opens the door to a plurality of methods, each in tune with a sphere of reality and a problem. On the other hand, in spite of the remarkable progresses made by scientists, we can find problems that there are issues that are not solvable by us in present times and even questions that might not be ever solved in any future by human beings (i.e. insolubilia).

Hence, (i) the structural complexity and the dynamic complexity of science does not allow, in principle, a method or a set of methods that can be universally valid for any kind of object (natural, social, and artificial). *De facto*, history of science has shown a spectacular advancement of very specialized methods in each scientific realm (in the natural sciences, in the social sciences, and in the sciences of the artificial). (ii) Regarding the future, the methodological difficulties posed by complexity to scientific research might have a somewhat similar status to scientific prediction in the following way. There are some events that might be "unpredictable" (i.e. a full impossibility for the humans) whereas other events are "not

---

for measurement of national economies, where social and economic elements are deeply interwoven.

[31] This involves the possibility of legitimate human knowledge outside science. This is stressed by Rescher; see Rescher (1999), chapter 4, pp. 99-121.

[32] This distinction has Kantian roots. A classical attempt to develop it is in Radnitzky (1978).

[33] Many years ago, Leszek Kołakowski made this remark in a conversation with me.

predictable" by human beings (i.e. we are not able to predict now). Thus, we can think of some complex phenomena that we will not be able to grasp ever by scientific methods, whereas there might be other complex phenomena that nowadays we are not able to grasp, but might be able in the future.

Given this methodological context on limits of science, we can think of the distinction proposed by Uskali Mäki between "economic expansionism," "economics imperialism," and "non-imperialism economics expansionism."[34] In his view, "economics expansionism is a matter of a persistent pursuit to increase the degree of unification provided by an economic theory by way of applying it to new types of phenomena" (Mäki 2009, p. 359). This is a well-known form of enlargement of scientific knowledge, which does not pose any relevant difficulty for a methodological universalism, insofar as it tries to combine intensity and expansion within a scientific realm.[35]

Meanwhile *economics imperialism* is characterized as "a form of economics expansionism where the new types of explanandum phenomena are located in territories that are occupied by disciplines other than economics" (Mäki 2009, p. 360). This is a kind of methodological universalism that, as mentioned before, fits quite well into the influential tradition of "economic imperialism" defended at least by economists of the Chicago school. This was accepted by George Stigler, who saw economics as an imperial science, insofar as "it has been aggressive in addressing central problems in a considerable number of neighboring social disciplines, and without any invitations" (Stigler 1984, p. 311).

For Stigler, this noticeable influence of economics on other fields can be seen in four territories: law, history, social structure and behavior, and politics. Thus, the methodological imperialism can be seen at least in several cases: 1) the economics of law, with the application of economic analysis to legal rules and legal institutions, is in Ronald Coase and

---

[34] He uses 'economics imperialism' instead of the usual denomination 'economic imperialism'. For him, the customary expression "denotes both the imperialism of the discipline of economics in the academic realm and the economy-driven imperialism in international relations and the global economy" (Mäki 2009, p. 352). Thus, he proposed using 'economics imperialism' to refer to the former and, thereby, to keep the two distinct. But it seems to me that the new expression does not solve the problem that he has pointed out. Nowadays, at least in the countries of the eurozone, both expressions might have the same meaning for the citizens.

[35] Besides the conceptions of scientific explanation as unification (cf. Gonzalez 2002), this perspective of "expansionism" seems that might be also accepted by methodological approaches developed by different philosophical tendencies (cf. Gonzalez 2006a).

Richard Posner; 2) the new history made in economic terms is in Robert Fogel[36]; 3) the economics analysis of social structure and behavior (crime, racial discrimination, divorce, etc.) is developed by Gary Becker; and 4) the economic analysis of politics, for example of constitutional design, is used by James Buchanan and Gordon Tullock, the founders of the "Public Choice" school.[37] In all of them, the repercussion of economics is on a relevant scale and with a large number of specialists.

Obviously, the problems posed by complexity in their several facets (structural and dynamic, epistemological and ontological) involve the existence of important methodological limits to this tradition of "economic imperialism." Moreover, Mäki himself proposes the possibility of a "non-imperialistic economics expansionism," which he conceives as "a form of economics expansionism where the new types of explanandum phenomena are located in unoccupied territories, that is, territories unoccupied by disciplines other than economics" (Mäki 2009, p. 360). This is a traditional way for the advancement of science, which is in tune with the ideal of a methodological universalism that comes from a specific science.

To sum up, there are several kinds of methodological universalism, which pertain to several levels (science, group of sciences, and specific sciences). These versions of methodological universalism can be conceived in terms of an "ideal" rather than a "goal." *De facto*, there are limits of science: on the one hand, the existence of "frontiers" or "barriers" of human knowledge; and, on the other, the possibility of a "ceiling" or a "final end" of the search made by this human undertaking (at least, regarding some problems). In this regard, there are sets of obstacles that come from complexity, both structural and dynamic, which involve epistemological and ontological elements.

They have a direct repercussion on what the methodological universalism can actually achieve. A facet of basic science and applied science where this is particularly noticeable is in the analysis of scientific prediction in economics, which is the discipline where "methodological imperialism" has received particular attention. The obstacles to scientific prediction emphasize the need for a scientific method to be in tune with an object (i.e. a sphere of reality) and a problem, which opens the door to a plurality of methods. In this regard, the methodological pluralism can lead

---

[36] On Fogel – Nobel Prize winner in 1993 – and the methodology of the "new history," see Gonzalez (1996), pp. 25-111; especially, pp. 29, 37, 74-75, 86, 90-91, 95, 105, and 107.

[37] Buchanan received the Nobel Laureate in economics in 1986. Regarding his methodological views, and in particular his approach on prediction in economics, see Gonzalez (2006b), pp. 89-90 and 100-101.

us to a different version of universalism in methodology of science: the analysis might show something that is shared by the *diversity of methods* used in science (natural, social, or the artificial).

University of A Coruña
Faculty of Humanities
Dr Vazquez Cabrera Street, w/n
15403-Ferrol (A Coruña)
Spain
e-mail: wencglez@udc.es

REFERENCES

Adriaans, P. and Van Benthem, J.K., eds. (2008). *Philosophy of Information*. Amsterdam: Elsevier.

Agazzi, E. ([1969] 1974). *Temi e problemi di Filosofia della Fisica*. Milan: Manfredi. 2nd ed.: (Roma: Abete).

Agazzi, E. and L. Montecucco, eds. (2002). *Complexity and Emergence*. Singapore: World Scientific.

Almeder, R. (2008). The Limits of Science, Realism, and Idealism. In: R. Almeder (ed.), *Rescher Studies. A Collection of Essays on the Philosophical Work of Nicholas Rescher*, pp. 1-28. Heusenstamm: Ontos Verlag.

Anderson, P.W., K.J. Arrow and D. Pines, eds. (1988). *The Economy as an Evolving Complex System*. Santa Fe, NM: Santa Fe Institute.

Ayer, A.J., ed. (1959). *Logical Positivism*. Chicago: Free Press of Glencoe.

Ayer, A.J. (1981). The Vienna Circle. *Midwest Studies in Philosophy* 6, 173-188.

Barkley Rosser Jr., J. (1999). On the Complexities of Complex Economic Dynamics. *Journal of Economic Perspectives* 13 (4), 169-192. Reprinted in: J. Barkley Rosser, Jr. (ed.), *Complexity in Economics*. Vol. 1: *Methodology, Interacting Agents and Microeconomic Models*, pp. 74-97. Cheltenham: E. Elgar. 2004.

Becker, G.S. (1976). *The Economic Approach to Human Behavior*. Chicago: The University of Chicago Press.

Becker, G.S. (1981). *A Treatise on the Family*. Cambridge, MA: Harvard University Press.

Bertuglia, Ch.S. and F. Vaio, eds. (2005). *Nonlinearity, Chaos and Complexity. The Dynamics of Natural and Social Systems*. Oxford: Oxford University Press.

Cabrillo, F. (1996). *The Economics of Family and Family Policy*. Cheltenham: E. Elgar.

Carnap, R. (1932). Die physikalische Sprache als Universalsprache der Wissenschaft. *Erkenntnis* 2, 432-465. English translation by Max Black in: R. Carnap, *The Unity of Science* (Bristol: Thoemmes Press, 1995).

Carnap, R. (1938-1955). Logical Foundations of the Unity of Science. In: O. Neurath, R. Carnap and C. W. Morris (eds.), *International Encyclopedia of Unified Science*, 1 (1) pp. 42-62. Chicago: The University of Chicago Press.

Chu, D., R. Strand and R. Fjelland (2003). Theories of Complexity. Common Denominators of Complex Systems. *Complexity* 8 (3), 19-30.

Clements, M. P. and D.F. Hendry (2002). Explaining Forecast Failure in Macroeconomics. In: M. Clements and D.F. Hendry (eds.), *A Companion to Economic Forecasting*, pp. 539-571. Oxford: Blackwell.

Dinga, E. (2009). Imperialism and Theoricity in the Economic Science. *Studii Financiare* **13** (1), 183-191.

Ericsson, N.R. (2002). Predictable Uncertainty in Economic Forecasting. In: M. Clements and D.F. Hendry (eds.), *A Companion to Economic Forecasting*, pp. 19-44. Oxford: Blackwell.

Franses, Ph.H. (2006). Forecasting in Marketing. In: G. Elliot, C.W.J. Granger and A. Timmerman (eds.), *Handbook of Economic Forecasting: Volume 1*, pp. 983-1012. Amsterdam: Elsevier.

Friedman, M. (1999). *Reconsidering Logical Positivism*. Cambridge: Cambridge University Press.

Giere, R.N. (1984). Toward a Unified Theory of Science. In: J.T. Cushing, C.F. Delaney and G.M. Gutting (eds.), *Science and Reality*, pp. 5-31. N. Dame: University of Notre Dame Press.

Gomez, A. (2011). New Forms of Scientific Observation and Their Epistemological Impact. In: W.J. Gonzalez (ed.), *New Methodological Perspectives on Observation and Experimentation in Science*, pp. 71-82. A Coruña: Netbiblo.

Gomez, A. (2010). Physics and Biology in the Social Science Method Debate. In: E. Agazzi and G. Di Bernardo (ed.), *Relations between natural Sciences and Human Sciences*, pp. 207-222. Genoa: Tilgher.

Gonzalez, W.J. (1994). Economic Prediction and Human Activity. An Analysis of Prediction in Economics from Action Theory. *Epistemologia* **17**, 253-294.

Gonzalez, W.J. (1996). Caracterización del objeto de la Ciencia de la Historia y bases de su configuración metodológica. In: W.J. Gonzalez (ed.), *Acción e Historia. El objeto de la Historia y la Teoría de la Acción*, pp. 25-111. A Coruña: Publicaciones Universidad de A Coruña.

Gonzalez, W.J. (1998). Prediction and Prescription in Economics: A Philosophical and Methodological Approach. *Theoria* **13** (32), 321-345.

Gonzalez, W.J., ed. (2002). *Diversidad de la explicación científica*. Barcelona: Ariel.

Gonzalez, W.J. (2003a). From *Erklären-Verstehen* to *Prediction-Understanding*: The Methodological Framework in Economics. In: M. Sintonen, P. Ylikoski and K. Miller (eds.), *Realism in Action: Essays in the Philosophy of Social Sciences*, pp. 33-50. Dordrecht: Kluwer.

Gonzalez, W. J. (2003b). Racionalidad y Economía: De la racionalidad de la Economía como Ciencia a la racionalidad de los agentes económicos. In: W. J. Gonzalez (ed.), *Racionalidad, historicidad y predicción en Herbert A. Simon*, pp. 65-96. A Coruña: Netbiblo.

Gonzalez, W.J. (2004). The Many Faces of Popper's Methodological Approach to Prediction. In: Ph. Catton and G. Macdonald (eds.), *Karl Popper: Critical Appraisals*, pp. 78-98. London: Routledge.

Gonzalez, W.J. (2006a). Novelty and Continuity in Philosophy and Methodology of Science. In: W.J. Gonzalez and J. Alcolea (eds.), *Contemporary Perspectives in Philosophy and Methodology of Science*, pp. 1-28. A Coruña: Netbiblo.

Gonzalez, W.J. (2006b). Prediction as Scientific Test of Economics. In: W.J. Gonzalez and J. Alcolea (eds.), *Contemporary Perspectives in Philosophy and Methodology of Science*, pp. 83-112. A Coruña: Netbiblo.

Gonzalez, W.J. (2008a). Evolutionism from a Contemporary Viewpoint: The Philosophical-Methodological Approach. In W.J. Gonzalez (ed.), *Evolutionism: Present Approaches*, pp. 3-59. A Coruña: Netbiblo.

Gonzalez, W.J. (2008b). Rationality and Prediction in the Sciences of the Artificial: Economics as a Design Science. In: M. C. Galavotti, R. Scazzieri and P. Suppes (eds.), *Reasoning, Rationality and Probability*, pp. 165-186. Stanford: CSLI Publications.

Gonzalez, W.J. (2011a). Complexity in Economics and Prediction: The Role of Parsimonious Factors. In: D. Dieks, W.J. Gonzalez, S. Hartman, F. Stadler, Th. Uebel and M. Weber (eds.), *Explanation, Prediction, and Confirmation*, pp. 319-330. Dordrecht: Springer.

Gonzalez, W.J. (2011b). Conceptual Changes and Scientific Diversity: The Role of Historicity. In: W.J. Gonzalez (ed.), *Conceptual Revolutions: From Cognitive Science to Medicine*, pp. 39-62. A Coruña: Netbiblo.

Gonzalez, W.J. (forthcoming). The Sciences of Design as Sciences of Complexity: The Dynamic Trait. In: D. Dieks, W.J. Gonzalez, Th. Uebel, M. Weber and G. Wheeler (eds.), *New Challenges to Philosophy of Science*. Dordrecht: Springer.

Granger, C.W.J. (2001). Evaluation of Forecasts. In: D.F. Hendry and N.R. Ericsson (eds.), *Understanding Economic Forecasts*, pp. 93-103. Cambridge, MA: The MIT Press.

Heylighen, F., J. Bollen and A. Riegler, eds. (2011). *The Evolution of Complexity*. Dordrecht: Springer.

Hofbauer, J. and K. Sigmund (1988). *The Theory of Evolution and Dynamical Systems*. Cambridge: Cambridge University Press.

Hodge, J. and G. Radick, eds. (2003). *The Cambridge Companion to Darwin*, Cambridge: Cambridge University Press.

Hoover, K.D. (2002). Econometrics and Reality. In: U. Mäki (ed.), *Fact and Fiction: Foundational Issues on Economics and the Economy*, pp. 152-177. Cambridge: Cambridge University Press.

Howe, L. (1993). Predicting the Future. In: L. Howe and A. Wain (eds.), *Predicting the Future*, pp. 1-7. Cambridge: Cambridge University Press.

Kuhn, Th.S. ([1962] 1970). *The Structure of Scientific Revolutions*, (International Encyclopedia of Unified Science: Foundations of the Unity of Science, vol. 2, n. 2). Chicago: The University of Chicago Press,.

Kuorikoski, J. and A. Lehtinen (2010). Economics Imperialism and Solution Concepts in Political Science. *Philosophy of the Social Sciences* **40** (3), 347-374.

Mainzer, K. ([1994] 2007). *Thinking in Complexity. The Computational Dynamics of Matter, Mind, and Mankind*. Berlin: Springer.

Mäki, U. (2009). Economics Imperialism: Concept and Constraints. *Philosophy of the Social Sciences* **39** (3), 351-380.

Mirowski, Ph. (1989). *More Heat than Light: Economics as Social Physics, Physics as Nature's Economics*. N. York: Cambridge University Press.

Mitchell, S.D. (2009). *Unsimple Truth: Science, Complexity, and Policy*. Chicago: The University of Chicago Press.

Morgan, M.S. (1990). *The History of Econometric Ideas*. Cambridge: Cambridge University Press.

Nelson, R. and S. Winter (1982). *An Evolutionary Theory of Economic Change*. Cambridge, MA: Harvard University Press.

Niiniluoto, I. (1984). *Is Science Progressive?* Dordrecht: Reidel.

Niiniluoto, I. (1993). The Aim and Structure of Applied Research. *Erkenntnis* **38**, 1-21.

Niiniluoto, I. (1995). Approximation in Applied Science. In: M. Kuokkanen (ed.), *Idealization VII: Structuralism, Idealization and Approximation* pp. 127-139. *Poznań Studies in the Philosophy of Sciences and the Humanities*, vol. 42. Amsterdam: Rodopi.

Parrini, P., W.C. Salmon and M.H. Salmon, eds. (2003). *Logical Empiricism: Historical and Contemporary Perspectives*. Pittsburgh: University of Pittsburgh Press.

Pies, I. and M. Leschke, eds. (1998). *Gary Beckers ökonomischer Imperialismus*. Tübingen: Mohr Siebeck.

Popper, K.R. ([1935] 1959). *Logik der Forschung*. Vienna: Julius Springer Verlag (reprinted by Tübingen: J.C.B. Mohr, P. Siebeck, 1994). English translation: as *The Logic of Scientific Discovery*. London: Hutchinson, 1959 (revised edition in 1968; reprinted by London: Routledge, 2001).

Radnitzky, G. (1978). The Boundaries of Science and Technology. In: *The Search for Absolute Values in a Changing World, Proceedings of the VIth International Conference on the Unity of Sciences*, vol. II, pp. 1007-1036. N. York: International Cultural Foundation Press.

Radnitzky, G. and P. Bernholz, eds. (1987). *Economic Imperialism: The Economic Method Applied Outside the Field of Economics*. N. York: Paragon House.

Rescher, N. (1998a). *Complexity: A Philosophical Overview*. New Brunswick, NJ: Transaction Publishers.

Rescher, N. (1998b). *Predicting the Future*. New York: State University Press New York.

Rescher, N. (1999). *Razón y valores en la Era científico-tecnológica*. Barcelona: Paidós.

Rescher, N. ([1984] 1999). *The Limits of Science*, revised edition. Pittsburgh: University of Pittsburgh Press.

Rescher, N. (2006). The Berlin School of Logical Empiricism and its Legacy. *Erkenntnis*, **64**, 281-304.

Simon, H.A. (1989). The State of Economic Science. In: W. Sichel (ed.), *The State of Economic Science. Views of Six Nobel Laureates*, pp. 97-110. Kalamazoo, MI: W. E. Upjohn Institute for Employment Research.

Simon, H.A. (1990). Prediction and Prescription in Systems Modeling. *Operations Research* **38**, 7-14.

Simon, H.A. ([1969] 1996). *The Sciences of the Artificial*. Cambridge, MA: The MIT Press.

Simon, H.A. (2000). Bounded Rationality in Social Science: Today and Tomorrow. *Mind and Society* **1** (1), 25-39.

Simon, H.A. (2001). Complex Systems: The Interplay of Organizations and Markets in Contemporary Society. *Computational and Mathematical Organization Theory* **7**, 78-85.

Simon, H.A. (2002). Near Decomposability and the Speed of Evolution. *Industrial and Corporate Change* **11** (3), 587-599.

Smith, P. (1998). *Explaining Chaos*. Cambridge: Cambridge University Press.

Stigler, G.J. (1984). Economics: The Imperial Science? *Scandinavian Journal of Economics* **86**, 301-313.

Strevens, M. (2003). *Bigger than Chaos: Understanding Complexity through Probability*. Cambridge, MA: Harvard University Press.

Taagepera, R. (2008). *Making Social Sciences More Scientific. The Need for Predictive Models*. Oxford: Oxford University Press.

Williamson, O.E. (1975). *Markets and Hierarchies: Analysis and Antitrust Implications*. N. York: Free Press.

Worrall, J. (2001). De la Matemática a la Ciencia: Continuidad y discontinuidad en el Pensamiento de Imre Lakatos. In: W.J. Gonzalez (ed.), *La Filosofía de Imre Lakatos: Evaluación de sus propuestas*, pp. 107-128. Madrid: UNED.

Eliza Karczyńska

ORIENTALISM AS A SIGN OF PROVINCIALISM

ABSTRACT. This article deals with various responses to the phenomenon of Orientalism. Since the publication of Edward Said's book *Orientalism*, there has been an ongoing discussion about the influence of Orientalism on contemporary social sciences in the East. In the West, Orientalism was an original theory, but in the East its acceptance was tantamount to an assimilation of foreign point of view on social reality. I argue that it is a symptom of provincialism among scientists from the East. Even though most of them tried to overcome Orientalism, they used the same categories and methodology. In this sense they repeated its mistakes and misunderstandings. This article analyzes different attempts of overcoming Orientalism and shows why they are provincial.

1. Introduction

Since the ancient times Europe has been interested in the East, trying to define its essence and specifics. Orientalism, stemming from these studies, was on one hand a group of sciences and on the other – a particular way of looking at countries like Persia, Ottoman Empire, and later also China, India or Japan.

The purpose of this article is to analyze the impact of Orientalism on contemporary social sciences in countries belonging to the Orient. Basic assumptions of Orientalism are still visible in many ideas of authors from East, even when they declare their separation from this concept. In the West, Orientalism was an original theory, but in the East its acceptance was related with assimilation of foreign view on social reality. I argue it is a symptom of provincialism among scientists from the East. The surprising thing is the fact that Orientalism was largely regarded as erroneous, distorted image of the East, which had to be overcome.

In the first part of this article, I will present the determinants of Orientalism and its social functions. I will refer here to Anouar Abdel-Malek and Edward Said's works, which gave an impulse to contemporary

In: Krzysztof Brzechczyn and Katarzyna Paprzycka (eds.), *Thinking about Provincialism in Thinking* (*Poznań Studies in the Philosophy of the Sciences and the Humanities*, vol. 100), pp. 177-195. Amsterdam/New York, NY: Rodopi, 2012.

discussions on Orientalism. In the second part I will show the impact of Orientalism on the modern state of science in countries belonging to the East. Finally, I will analyze attempts of overcoming Orientalism and show why they are provincial.

2. On Two Dimensions of Provincialism in Science

The phenomenon of provincialism in science can be viewed in two dimensions: intellectual and socio-economic. An example of the first one is an analysis of provincialism developed by Leszek Nowak, the second one can be seen in the works of Johan Galtung.

In his paper "The Structure of Provincial Thought" ([1998] 2012), Leszek Nowak analyzes the intellectual dimension of provincialism in science. From this point of view, a scientist is less important than the environment, in which he operates:

> This means that the mere fact of belonging to a given research community in a given science is taken to be a sign of the cognitive role an individual is taken to play in that science. If you are from the center, you should create, for you have the right to do so. If you are from a province, you should know better not to be a smart aleck and apply what has been created in the center. If you are from the gray area, you can try to improve (but not change!) "central theories." (Nowak [1998] 2012, p. 63)

Thus a division emerges, according to which some centers aspire to a creative role of being a source of new theories. Others are assigned the role of correctors who introduce modifications or extensions to the existing theories. Finally, there are centers which serve as applicators. Their role is to use existing theories to explain new issues (Nowak 1998, p. 16). If they try to ask new questions, they are only such to which the existing theory has an answer.

A hierarchy is thus formed, where some centers have a dominant influence, while others are subordinated. Provincialism is based on this notion of inferiority. It manifests itself in applying a mindset founded in the center by the less significant (or simply younger) scientists. The provincial authors recognize theories coming from the center as more valuable. You can see it in the direction of citation:

> The point is not just the mere direction of citations but the fact that it depicts the hierarchy of world-widely accepted cognitive roles. However, the hierarchy of accepted cognitive roles can (and perhaps even usually does), but certainly need not, replicate the hierarchy of actual cognitive

achievements. This is evident if only from the fact that the former is preordained. (Nowak [1998] 2012, p. 63)

For the center, such a situation appears to be favorable while, for the province, it is related to the dependence and suppression of independent research. The province has the role of an applicator, at best of the corrector. Hence, in case of creation of a new original theory, researchers from these centers are struggling to spread their ideas.

It is assumed that the centers that are the most creative and original have the largest impact. The intellectual provincialism in science is based on this assumption. Beliefs of the scientific community about the status and the potential of these centers seem to be crucial. Research conducted at major centers is valued more on the basis of its origin. Meanwhile, the impact of a particular concept or institution may also depend on other factors. It seems that socio-geographic factors are important as well. One indicates a geographical division of the world into the developed countries and the rest of the world. Political and economic dominance on this approach would have an impact on inequalities within science. Of course, it is impossible to deny that more funds for research, publications, development of technology have an influence on the status of individual centers. Globally, stronger centers impose their way of thinking and performing science. Centers operating within the politically and culturally dominant countries create research centers, which are more powerful than the peripheries, regardless of the theories they create within. There is a hierarchy, which does not necessarily reflect the purely scientific categories. This division is then translated into subordinate relations between research centers operating in the center and the periphery. The knowledge created in the peripheral areas has to match the knowledge formed in the center in order to be regarded as legitimate.

This brings us to the socio-economic dimension of provincialism in science. What is important here is the influence of social, political and economic relations on the development of science. Division of the world into centers and peripheries was the basis of Johan Galtung's concept of imperialism. Galtung distinguished five types of imperialism: economic, political, military, communicative and cultural (Galtung 1971, p. 91). Within cultural imperialism we can find scientific imperialism which is characterized by the fact that "the center of attraction of the acquisition of knowledge is located outside the nation" (Galtung 1967, p. 296). Whereas Nowak divides society of scientists into creators, correctors and applicators, Galtung introduces a division into teachers and learners. However, it is not this division that is the most significant for scientific imperialism but the location of these two groups. The center provides teachers who point out what is worth learning. The periphery is the source of the learners. All theories coming from peripheries cannot be at the same level as the ones from the centers because it would distort the hierarchy. Galtung argues that peripheries are more responsible than the center for maintaining that state of fact. Of course, maintaining this hierarchy seems natural for the center. But peripheries, hoping for such benefits as new technologies or opportunities for learning, strengthen this division further. The position of the center is getting stronger while periphery stops to produce any new original theories which would undermine the competence of the center.

Nowadays the phenomenon of provincialism is usually related to the American dominance in the world. However, at the beginning of the 20th century, the United States were considered to be provincial. In the essay *Provinciality* (1940), Joseph Baker defined provincialism as a "mental state produced by living in a 'region dependent on a distant authority'" (Baker 1940, p. 489). He criticized cultural dependence of the United States on Europe, as well as their internal provincialism:

> And what is true of the relation between Europe and America is true of the relation between our North Atlantic seaboard and the rest of the United States. Indeed, it is worse. [. . .] Academics in the hinterland are for the most part timid and deferential toward the eastern colleges, often to the point of betraying the very standards that made the older colleges great: they will sub-ordinate their own better judgment to eastern degrees and eastern customs, making recognition depend on eastern reputations, even when the superior worth of the local work is evident. Literary clubs buy the books recommended by eastern reviews. (Baker 1940, p. 942)

The fact, that the United States managed to get rid of this provincial trait did not mean the disappearance of this phenomenon. However, its scope has changed. Whereas earlier provincialism was limited to countries within European culture, now it relates to a much wider area.

3. On Orientalism

The division of the world into the West (Occident) and the East (Orient) goes back to ancient times. In 1314, Church Council in Vienna decided to establish chairs in Greek, Arabic, Hebrew and Syriac at universities of Paris, Avignon, Bologna, Oxford and Salamanca (Macfie 2000, p. 19). In 17th and 18th centuries overseas journeys and missionary work of Jesuits were an impulse to founding research institutes across Europe dealing with language and culture of India, China and Japan. As an independent academic discipline Orientalism developed in France and Great Britain in the 19th century.

According to A.L. Macfie, three notions of Orientalism existed by the end of the Second World War. First, an orientalist was a researcher dealing with the culture and the languages of the Orient (Turkey, Syria, Mesopotamia, Arabia, Palestine, later also India, Japan, China, and even the whole Asia). Second, Orientalism denoted a characteristic style of arts in these countries. Third, it referred to the "conservative and romantic" approach that characterized the way of governing of the representatives of Great Britain in East India Company (Macfie 2000, p. 1). Orientalism acquired a new dimension after the Second World War, when it began to be perceived as a certain set of beliefs of the Western scholars relating to the East. It started a new reflection on social impact of Orientalism.

Edward Said's book *Orientalism* is considered to be the flagship text relating to a reflection on the essence of Orientalism. The work still inspires much discussion. However, Said was not the first who undertook a critique of Orientalism. An equally important, though currently often overlooked, text is an article "Orientalism in Crisis" by Anouar Abdel-Malek, an Egyptian sociologist working in Paris.

4. The Essence of Orientalism

Although Orientalism as a study dedicated to the Orient has a centuries-old history, a reflection on its essence and social influence appeared in the second half of the 20th century. The main reason was the emergence of new states on the ruins of the old colonial system. One of the first scholars who took up this theme was Anouar Abdel-Malek. In his article "Orientalism in Crisis" (1963), he put forward a critique of the historical premises of Orientalism. Abdel-Malek focused on distinguishing the characteristics of studies of the Orient, creating a static image of this discipline. Orientalism was to use a specific methodology and to be based on some very general assumptions. In terms of the subject matter, Orient was the *object* of research characterized as being different. As a passive object, it was non-autonomous, unable to act. Hence, it could only be defined, understood by others. In terms of content, orientalists accepted essentialist conception of countries and people of the Orient, which expressed itself through an ethnic typology:

> According to the traditional orientalists, an essence should exist – sometimes even clearly described in metaphysical term – which constitutes the inalienable and common basis of all the beings considered; this essence is both 'historical', since it goes back to the dawn of history, and fundamentally a-historical, since it transfixes the being, 'the object' of study, within its inalienable and non-evolutive specificity, instead of

defining it as all other beings, states, nations, peoples and cultures – as a product, a resultant of the vection of the forces operating in the field of historical evolution. (Abdel-Malek 1963, p. 108)

Unlike the Europeans, whose essence was shaped through history, it was possible to define *homo Sinicus, homo Arabicus* or *homo Africanus who remained* resistant to any changes.

Among Western scholars, there was an assumption that the Orient countries had the best time behind them and were now in the phase of decline and decadence. Therefore, to capture their accurate essence, it was necessary to study the past, not the present. The majority of these studies focused on dead languages. Their field was limited to the language and culture and, at the same time, it ignored the social aspects. This separation from the social context had a consequence in the acquiring of irrational features by the Orient. The description of reality was not important; what really mattered were some general ideas. The way in which orientalists conducted their research was symptomatic. They ignored achievements of oriental thinkers, which in turn formed the belief that the countries of the Orient were devoid of native social reflection.

Abdel-Malek also pointed to the impact of imperial politics on the access to the materials. The vast amount of valuable materials was in the possession of imperial authorities. This caused obstacles encountered by researchers from Eastern countries after gaining independence. It appeared that researchers from Egypt or India had less access to source materials than their counterparts from Great Britain or France. For example, the vast part of the British collection was subject to various kind of interdictions (for example for fifty years) and was not accessible to researchers from the East. Due to that fact they had to use indirect sources such as diaries and reports from expeditions, often imbued with racist and ethnic prejudices (Abdel-Malek 1963, p. 111). Indirect sources were written by researchers working on behalf of the imperial governments or by missionaries. Using the formulation of Galtung, it was an example of scientific imperialism. The authorities of the metropolis imported cultural goods (as they did in case of raw materials) from subordinate colony without any problems. Then they produced finished products (articles, books), which were sent back to countries of the East, where they were becoming the object of analysis. The center of learning thus changed its location.

Although "Orientalism in Crisis" met with positive response from researchers from the Orient, it did not provoke wider discussion. That happened in the end of the 1970s. The book by Edward Said's *Orientalism* (1977) became an impulse for the debate. The Palestinian intellectual distinguished several uses of the word 'Orientalism'. In his opinion, the most important one was connected with the power. Referring to Foucault,

Said described Orientalism as a discourse based on the relation of dominance and strength:

> [Orientalism] is, above all, a discourse that is by no means in direct, corresponding relationship with political power in the raw, but rather is produced and exists in an uneven exchange with various kinds of power, shaped to a degree by the exchange with power political (as with a colonial or imperial establishment), power intellectual (as with reigning sciences like comparative linguistics or anatomy, or any of the modern policy sciences), power cultural (as with orthodoxies and canons of taste, texts, values), power moral (as with ideas about what "we" do and what "they" cannot do or understand as "we" do). (Said 1977, p. 12)

Orientalism in a broader sense was a way of thinking based on the ontological and epistemological distinction between "the East" and "the West." In a narrower sense, Orientalism meant a scientific discipline.

We can distinguish several determinants in Said's concept. It is important to keep in mind that Said consciously limited his study to the analysis of British, French and American influences. That fact had significant implications for the relationship between Orientalism and imperialism (Said 1977, p. 15). The second assumption adopted by Said was the monolinearity of history. Although he did not write so explicitly, Said regarded history as a narration, coherent and uniform for all societies. Hence, in his book, all the elements of the puzzle called Orientalism match one another. Although the author focuses primarily on the 19th and first half of the 20th century, at the end of his book, while analyzing contemporary research of the Orientalists, he repeats the main assumptions of his previous studies. He concludes that the approach of the Western scholars to the Orient did not change significantly. According to Said, Orientalism was characterized in terms of five features.

a) *Incapacity of self-representation.* Said began his book with a quote from Karl Marx: "They cannot represent themselves, they must be represented" (Said 1977, p.1). This sentence, originally referring to French villagers, was the starting point of Orientalism, according to the writer. You can see here Abdel-Malek's idea that the East was treated as an *object* of study, not a *subject* that would be equivalent to the West. Something that is merely an object cannot describe on its own. If the Orient could not represent itself, it was naturally put on a lower level than those who were able to describe themselves. This was a fundamental assumption since any possibility of knowledge coming from the Orient was thus negated. The knowledge of the Orient had to come from the outside. Furthermore, the tools and theories created in the West were considered to be adequate (because the only possible) to explore the people of the Orient. For Said, the chief issue was representation, although he expressed doubts about the accurate representation of a phenomenon:

> [. . .] the real issue is whether indeed there can be a true representation of anything, or whether any and all representations, because they *are* representations, are embedded first in the language and then in the culture, institutions, and political ambience of the representer. If the latter alternative is the correct one (as I believe it is), then we must be prepared to accept the fact that a representation is *eo ipso* implicated, intertwined, embedded, interwoven with a great many other things besides the "truth," which is itself a representation. (Said 1977, p. 273)

b) *Monolithicity*. Said argued that Orientalism had produced a coherent vision of the Orient.[1] These same features were used in determining the people of India, Egypt and the Ottoman Empire. This claim later became the object of sharp criticism. In his book, Said focused on researchers working within the structures of imperial Britain, France and the United States, which had similar goals and hence similar visions of the East. He ignored many eminent German scholars who traveled across the Orient without any aspirations of dominating it.

c) *Invariability*. According to Said, people of the West shared a common belief in the immutability of the Orient. It was possible to learn about the contemporary Orient by studying its ancient texts. While this assumption was widely rejected in the second half of the 20th century, abstractions and generalizations were still preferred over empirical evidence.

d) *Otherness*. It became possible to provide absolute, systematic differences between the West, which was rational, developed, humanistic, and the Orient, which was not typical, undeveloped, and worse. However, such representation treated not only the East but also the West as ahistorical. The assumption of rationality (and humanism) of the West would have its origin in ancient Greece. No attempts were made to find common characteristics of the East and the West since they were

---

[1] Farris claims, that there are similarities between Said's discourse and Weber's ideal type: "*Orientalism* shares with Weber's notion of the ideal type the one-sided viewpoint of departure and focus that implies the choice of what aspects or objects can enter into the definition (or class). Thus, Said chose to focus on British, French and American Orientalism, while leaving aside other writings. Like Weber's ideal type, *Orientalism* tries to view with the same 'telescope' centuries of history and the most disparate writings and authors, thus often risking annihilating history and dehistoricizing the object of investigation" (Farris 2010, p. 280).

ontologically opposite.[2] This had profound implications in the development of science about the Orient. The historical development of Orientalism as a field of research underwent expansion in its geographical scope. Science itself did not undergo greater specialization.

e) *Threat.* According to Said, in case of both traditional and contemporary Orientalism, it was possible to distinguish two trends. Orient was either to be feared ("yellow danger," Mongol hordes) or to be controlled (by pacification, research and direct control where possible).

5. Orientalism as a Sign of Provincialism

Orientalism among Western scholars is not a sign of provincialism but its adoption by the scholars from the East is. Paradoxically, the Eastern researchers, in their attempts to overcome Orientalism, often strengthen it.

In his book, Said presented a picture of an orientalist-researcher trying to explore the Orient objectively, in fact acting on behalf of political power. The main thesis of the Palestinian researcher was the claim that knowledge enables power: a better understanding of the society means better control. Orientalism was an instrument of political domination. On the other hand, knowledge as power seemed to be a natural consequence of the inability of the Orient to create its own self-representation. Since it was "weaker," it seemed obvious that it had to be controlled.

The main assumptions of Orientalism had a significant influence on contemporary development of scientific research in the Arab world:

> In the one part of the Orient that I can speak about with some direct knowledge, the accommodation between the intellectual class and the new imperialism might very well be accounted one of the special triumphs of Orientalism. The Arab world today is an intellectual, political, and cultural satellite of the United States. This is not in itself something to be lamented; the specific form of the satellite relationship, however, is. (Said 1977, p. 322)

---

[2] As a representation of the Other, Orientalism was used as well in order to criticize European mentality. According to Clarke: „During the Enlightenment period, for example, representations of foreign societies and tribes of all kinds, real and imaginary, were commonly deployed to criticise the follies and inadequacies of European civilisation; the myth of the Noble Savage is a well-known example of this genre. Thus Amy Glassner Gordon, in a discussion of French Enlightenment attitudes towards other cultures in general, points out that the philosophers 'believed that a wider perspective, a more thorough understanding of non-European societies, would enable them better to understand themselves and the world in which they lived'" (Clarke 1997, p. 28).

Said deplores the fact that "no Arab or Islamic scholar can afford to ignore what goes on in scholarly journals, institutes, and universities in the United States and Europe; the converse is not true" (Said 1977, p. 324). Orientalism has become a kind of extension of political domination and strengthened the divisions within the scientific community.

> One of the major themes running through Said's work is the distortion that power relations between the West and the Orient brings to scholarship; scholars of the powerful nations instinctively treat the history, and the present, of less powerful peoples as inferior, childish, at a lower stage of development. These peoples are the Other. Orientalist views derive from a sense of superiority over the Other; they then reinforce that sense of superiority by providing "evidence" of the inferiority of the Other. (Lary 2006, p. 4)

The critique of Orientalism was to abolish this division. It was mainly based on the belief that, since this division was imposed by force and was somewhat artificial for the countries of the Orient, it would now without any problems be rejected. However, both external (having its source in Western centers) and internal conditions made this inequality operating until now.

6. Extensions of Said's Analyses.
Intellectual Imperialism and the Captive Mind

Said's book provoked two kinds of reactions. On one hand, there was a deepening of his analyzes by Eastern researchers and, on the other hand, there was an attempt to overcome the vision of the world constructed in *Orientalism*. Both of these attitudes, albeit in different ways, were in fact manifestations of provincialism.

At the beginning, *Orientalism* was taken as a call to reject the hitherto dominant paradigm in studies devoted to the East. Optimists argued that, with the critique of Orientalism, it became possible to create more reliable research of the Orient that would be free from Eurocentric (exhibited by Eastern scholars) discourse (Teitelbaum and Litvak 2006, p. 24). This did not happen, however. The process of decolonization and the development of academic centers in the countries of the Orient did not mean the rejection of Orientalism. It turned out that the scientists of these countries have taken over most of the determinants of Orientalism distinguished by Said and began to apply them in their studies. Only some accents shifted. Said spoke about the use of Orientalism for political and economic domination. Now we can see similar phenomenon in the field of science. The mechanism, however, remained the same.

According to Syed Hussein Alatas, we can see parallels between the economic dominance and the state of science in the countries of the East. Intellectual imperialism, domination of one way of thinking, functions like economic imperialism. During the colonial period, metropolis treated subordinate territories as a source of raw materials. Finished products were exported back to the colony. A similar mechanism works nowadays in intellectual imperialism.

> The data from this region, the raw data on specific topics are collected and subsequently processed and manufactured in England in the form of books and articles, and finally sold here. (S.H. Alatas 2000, p. 25)

The researchers from the East thus became a source of information but not of theories or concepts. Their role is reduced to the transmission of data. Moreover, there is a change in perception of the image of science. The basic value is the accurate transmission of information while the creative interpretation of the data is attributed to the scientists of the West. For example, it is visible in the way that lecturers are employed at Turkish universities. Those researchers who spent at least part of their career in the centers of the West (mainly in the United States, Great Britain, France) are considered to be more creative and competent, even if the subject of their research is connected to the history or culture of Turkey.

Treating scholars from the East as merely a source of data was practiced in Western research centers as well. During the Cold War, China was a significant subject of study to the United States. The isolation of the Asian country meant that most of the studies had to be carried out outside of its territory. Thus, during the period from 1950s to 1980s, many American universities employed highly qualified Chinese scientists, who were predominantly political refugees, as assistants. They were regarded as specialists in the field of language and culture, and supplied raw data, on which more prominent American researchers conducted their analyses and formulated their theories. The contribution of Chinese scientists was very rarely shown in the final product. This was subsequently reflected in the scientific careers of immigrants from China. Since they were not treated as creative, independent individuals, usually they could not get promoted to high positions (Lary 2006, p. 6).

There is another feature related to intellectual imperialism. The belief in the backwardness and lower value of the Orient became a pretext for the West to take a role of a tutor. You can also find a reference here to economic imperialism. Great Britain and France were not willing to get rid of their colonies and justified it with the claim that these areas were not mature enough to reach independence. Similarly, in case of science, countries of the Orient are believed to be underdeveloped in education and research. If you wanted to gain objective knowledge, it was necessary to turn to the West.

> It was assumed that people here know less about practically all subjects than people in the West. Once again a parallel exists. In the past the outlook was that the colonies could not maintain themselves. They could not be granted independence because they would ruin the country if they govern themselves. They could not be relied upon to develop the country because they did not have the technical know-how. Now, the parallel with intellectual imperialism is that they do not have the intellectual know-how. Hence the need for a form of indirect tutelage. (S.H. Alatas 2000, p. 25)

The role of a tutor resulted not only in unequal division of knowledge (the tutor has a much higher degree of knowledge than the person who is under his tutelage) but also in a moral commitment to leadership.

Intellectual imperialism produces a way of thinking, which Alatas calls "the Captive Mind." While intellectual imperialism is imposed from the outside and represents a global approach, the Captive Mind is self-imposed. How is it characterized? Alatas (2000, p. 37) lists ten characteristics:

1. It is a product of higher institutions of learning, at home or abroad, whose way of thinking imitates, and is dominated by, Western thought in an uncritical manner.
2. It is uncreative and incapable of raising original problems.
3. Its method of thinking imitates, and is dominated by, Western thought in an uncritical manner.
4. It is incapable of separating the particular from the universal and consequently fails to adapt the universally valid corpus of knowledge to the particular local situations.
5. It is fragmented in outlook.
6. It is alienated from the major issues of society.
7. It is separated from its own intellectual pursuit.
8. It is unconscious of its own captivity and its conditioning factors.
9. It cannot be studied in a quantitative manner but can be studied through empirical observation.
10. It is the result of Western dominance upon the rest of the world.

The Captive Mind is provincial in the sense that it is unable to function independently, in case of social sciences it is unable to develop and create new theories without constantly looking up to the West. The consistent decline of native intellectual tradition puts scientists in a vacuum, from which the only escape seems to be following the European thought.

The lack of originality is Alatas' main charge against Eastern scholars. A partial explanation for this situation may be the fact that the social sciences have been imported from outside in a developed form. As a result, there is a lack of communication between European knowledge and indigenous ideas and history. Meanwhile, it is not true that there was no social science in the Orient before the era of colonialism. Syed Farid Alatas cites the example of the Arab thinker Ibn Khaldun living in the second half of the fourteenth century. In his numerous works, he wrote *inter alia* about the emergence of civilization, political power and many other issues. He distinguished two levels of knowledge in the historical sciences and created foundations for the development of social sciences. Unfortunately, nowadays his thought is ignored in studies. For example, there are no attempts to integrate the ideas of Ibn Khaldun with Western theories. In the countries of the Orient, we can observe a general denial of local history and of philosophical traditions. This is reminiscent of Abdel-Malek's claim that the world of the Orient was an object of research. The native tradition of Ibn Khaldun is merely an object of study, not a source of inspiration or philosophical concepts. This argument is also an extension of Said's idea about the Orient's inability to create its own representation. In the past, Orientalists ignored local traditions. Now things are only slightly better: these traditions are an object of exploration but not a source of knowledge.

There is another cause of intellectual imperialism. Funds, technology and prestige are unevenly distributed in the world of science. In addition, we may talk about a global market of scientific ideas. In this market, the most competitive ideas win, and they are determined by rhetorical programs. The spread of new vocabulary and terminology results in the spread of new ideas. When we learn new terminology, we absorb particular ways of thinking and looking at the world.

According to some researchers, imperialism caused a situation where we live in the world of one way of thinking, and "anything in the modern history of Asia that can be achieved from the perspective of Asia-centric, can be equally obtained by Western scholars" (Smail 1961, pp. 75-76). This belief in the superiority of science, which stems from the European tradition, limits the development in other areas of the world. It comes from the erroneous belief that outside Europe social sciences did not develop. In addition, it is still assumed that European thought is universal. Thus, research methods and theories seem to be adequate to study other traditions and cultures. N.K. Singh cites the example of sociology and the structural-functionalist approach, which dominated this field for a long time. Parson's theory of actions has become dominant for social scientists in India for a certain period. However, the model proposed by Parsons' action does not take into account the transcendental spiritual motivation. It ignores specific cultural factors in Indian society.

7. Attempts at Overcoming Orientalism.
Orientalism in Reverse and Occidentalism

The situation of intellectual dependence met with hostility and resistance at times. It was considered that since Orientalism was a foreign way of thinking imposed by force, it was necessary to reject it and build a new and a more appropriate discourse. However, most of these attempts derive from the methodological and ontological assumptions of Orientalism. The idea of Orientalism in Reverse was developed by the Syrian philosopher Sadiq Jalal al-'Azm:

> In classic Orientalist manner, the essence of the "Arab mind" is discovered by an Arab thinker only through language, in hermeneutical isolation from unwanted intruders such as socio-economic infrastructure, politics, historical change, conflict of classes, revolutions, etc. The original Arab mind, psyche whether the entity has to disclose its potency, and characteristics of genius by the stream of historical events. (Sadiq Jalal al-'Azm 2000, p. 232)

Orientalism in Reverse is based mainly on the analysis of texts, trying to extract from them the essence of the Arab mind. It is thus similar to classical Orientalism. It developed a specific view of the colonial period:

> If Orientalism had developed itself into Western ethnocentric, we observed that the former colonized people had now developed a reversed form – that of asserting their own culture and traditions as inherently superior to the West, or what Said termed as *nativism* [. . .] Similarly, we observed the same tendency amongst Muslims who seek to romanticize the past glories of Muslim civilization. Facing Western superiority politically, scientifically, economically and technologically, several Muslim thinkers wrote extensively on how the West "borrowed" their science and learning from the Muslim world. The Muslim world was once superior to the West and had produced some of the pioneering scientists like Ibn Sina (Avicenne), al-Hayatham and al-Razi. (Taib 2003, p. 5)

An attempt to formulate new discourse specific to the Arab world was in fact replicating the Orientalist way of thinking. Orientalism made the mistake to assume that there is a fixed and unchanging essence of the Orient. Meanwhile, similar assumptions, this time relating to the Western world, can be found in the works of many Muslim thinkers like Syed Qutb, Muhammad Asad, Maryan Jameelah. The West is a monolithic entity and is characterized by a set of characteristics. It is: (1) corrupt, (2) atheist, (3) conspiring, (4) materialistic, (5) immoral,[3] (6) bad, (7) anti-religious. In the past, there was a conviction that the West, as more developed, should serve as a teacher and mentor to the East. According to Orientalism in Reverse, since there is a decline of the West, the Arab world should take the *mission civilisatrice*.

Orientalism in Reverse is based on essentialist vision of both the Orient and the Western world. It is thus a kind of auto-Orientalism. A good example can be *nihonjinron*, a set of theories about Japan operating in this country. The society is viewed in an essentialist way. Its cultural homogeneity and historical continuity are emphasized. The only difference is that the subjects it explores are the Japanese, and not someone from outside.

Similar assumptions and methodology of formulating theories can be seen in Occidentalism, developed by a journalist Ian Buruma (the name is a deliberate reference to Said's *Orientalism*). This term describes the Asians who look down on the West, believing that everything that comes from Asia is better. This way of thinking combines nationalism with a reaction to Western imperialism and colonialism. In the book *Occidentalism*, written by Buruma together with Avishai Margalit, they argue that Occidentalism brings even greater danger than Orientalism:

> Some Orientalist prejudices made non-Western people seem less than fully adult human beings. . . . Occidentalism is at least as reductive; its bigotry simply turns the Orientalist view upside down. To diminish an entire society or a civilization to a mass of soulless, decadent, money-grubbing, rootless, faithless, unfeeling parasites is a form of intellectual destruction. (Buruma and Margalit 2004, p. 10)

According to Buruma and Margalit, Occidentalism as an idea of the West is based on four "archetypal forms." These are: the city, the Western mind, the settled bourgeoisie and the infidel. They all appeared as an expression of anti-Enlightenment tendencies and spread to various parts of the world, for example Japan of the 1940s and the Islamists nowadays (El-

---

[3] "The West is portrayed as corrupt, degenerate, uncaring and hypocritical. Beijing and Washington practice a tit-for-tat battle over human rights. Every time Washington cites violations in China, Beijing comes back with examples of inhumane treatment of disadvantaged people in the United States. China is particularly critical of the West's supposed lack of social responsibility. One of the Chinese values most strongly promoted is respect for the elderly. Old peoples' care homes are considered an anathema, a way of disposing of the elderly, not giving them the respect they deserve at the end of a long life" (Lary 2006, p. 10).

Haj 2005, p. 542). Buruma argues that the origin of Occidentalism was not in the East but on the European soil. It was a reaction to Enlightenment and industrialization (Buruma 2004, p. 2).

Orientalism was based on the belief that only Westerners were able to get to know the Orient objectively. Occidentalism reverses this assumption by claiming that no one but the Japanese are truly able to get to know Japan, no one but the Chinese can really get to know China, etc. No foreigners, regardless of their knowledge of culture, language or history of the country, can understand the culture, in which they were not born and raised. A consequence of Occidentalism is the belief that the East does not owe anything to the West except for technological innovation and Christianity. For example, China is trying to replace the Marxist values with native alternatives (that fulfill the same roles but on different grounds). Where the West has a rule of law *fazhi*, China has a rule of an individual-*renzhi* (Lary 2006, p. 10).

Both Orientalism in Reverse and Occidentalism are strongly influenced by Orientalism, and thus do not meet the expected role. They do not constitute new alternative discourses. However, there are attempts to build new ways of thinking. The most common postulate is to turn to the local culture. Syed Farid Alatas talks about the need for the formulation of alternative discourses.[4] They would look on indigenous historical experiences and practices in the same way as in case of the Western thought. It is necessary to open oneself to the native philosophers, epistemology, history, art and other forms of human knowledge (S.F. Alatas 2000, p. 5).

Similar assumptions are postulated by researchers from the circle of Subaltern Studies. They maintain that, in many countries of the Orient (mainly India), there were other subjects of social change than in the West. (Hence the name 'Subaltern', which refers to the lowest strata of society as to the subject of change.) For this reason, it is impossible to apply here European theories of social development.

The attempt to overcome the belief in the universality of the Western thought is the main postulate of Subaltern Studies scholars. According to Chakrabarty, Europe remains a theoretical sovereign subject of historical discourse and other cultures create their discourses by reference to it. All

---

[4] Alatas gives a definition of alternative discourses: "The term 'alternative discourses', therefore, is one that we are introducing and which should be understood as a descriptive and collective term referring to that set of discourses that had emerged in opposition to what they understand to be mainstream, Euroamerican social science. Alternative discourses constitute a revolt against 'intellectual imperialism'" (Alatas 2001, p. 58).

the other histories are thus the successive versions of "the history of Europe." In his opinion, this is a consequence of the fact that that European philosophers for centuries formed their theories, which were embraced by all mankind. However, these thinkers had knowledge only about a very small part of the world. Chakrabarty calls for the provincializing of Europe, and for treating its history not as a dominant one but rather as one of possible versions.

In the book *Culture and Imperialism*, Said presents another way to get rid of Orientalistic thinking. He does not question the impact of the colonial past on the countries of the Orient. However, in this case, their knowledge and analyses should be carried in parallel to the history of the metropolis – contrapuntally:

> An example of the new knowledge would be the study of Orientalism or Africanism and, to take a related set, the study of Englishness or Frenchness. These identities are today analyzed not as god-given essences, but as results of collaboration between African history and the study of Africa in England, for instance, or between the study of French history and the reorganization of knowledge during the First Empire. In an important sense, we are dealing with the formation of cultural identities understood not as essentializations (although part of their enduring appeal is that they seem and are considered to be like essentializations) but as contrapuntal ensembles, for it is the case that no identity can ever exist by itself and without an array of opposites, negatives, oppositions. (Said 1993, p. 52)

Conclusion

If it became possible to reject the unity of narration in describing the East and other concepts of history became visible, why do we have a phenomenon of provincialism among researchers coming from the Orient? It seems that the debate triggered by Said's work would start the creation of new independent discourses among Eastern scholars. Meanwhile, many of them did not dare to be independent and fell into provincialism.

The reasons for that can be found both in the socio-economic and intellectual dimension of provincialism. According to Clarke, we can reject the assumption of the unity of history but we can not get rid of the beliefs about the relationships of Orientalism with the authority and the resulting subordination.

> The history of the cultural and intellectual relationship between Europe and Asia over the past four centuries must inevitably be viewed in the light of the growth of Western military and economic power. Some have gone so far as to suggest that Orientalism, in its study of religious and

philosophical traditions of Asia, was merely the intellectual wing of the whole coordinated strategy of mapping, measuring, and classifying the peoples of Asia for the purpose of more efficient control, and that the role of orientalists was little different from that of the geographers and surveyors who were part of the colonial retinue. (Clarke 1997, pp. 25-26)

Samir Amin, an Egyptian sociologist and political scientist, connects the development of Orientalism with the creation of the capitalist system in Europe. The desire to conquer new territories and to achieve even higher profits meant that it became natural to evaluate the conquered peoples and to represent them. Since other countries were undeveloped (from the perspective of European lineage), they also could not have specialized tools of description. Polarization in the distribution of means, characteristic of capitalism resulted in a division of the world into the center and peripheries. Europe's own economic development has become a cause of dichotomous division of the world. This in turn had an impact on the image of science.

If we add Galtung's theory on five interrelated forms of imperialism to these two visions of development, we have the answer to the question of provincialism among researchers in the East. Scientific imperialism is one of the forms of power – dependent on the economy, politics, military and communication.

On the other hand, scientists from the East refused to adopt a creative attitude for a long time. Accustomed to the role of correctors or applicators, they have perpetuated this model.

Adam Mickiewicz University
Instytut Filozofii
ul. Szamarzewskiego 89C
60-568 Poznań
Poland
e-mail: eliza.karczynska@hotmail.com

REFERENCES

Abdel-Malek, A. (1963). Orientalism in Crisis. *Diogenes* **11**, 103-140.
Alatas, S.F. (2006). Ibn Khaldun and Contemporary Sociology. *International Sociology* **21** (6), 782-795.
Alatas, S.F. (2000). An Introduction to the Idea of Alternative Discourses. *Southeast Asian Journal of Social Sciences* **28** (1), 1-12.
Alatas, S.F. (2001). Alternative Discourses in Southeast Asia. *Sari* **19**, 49-67.

Alatas, S.H. (2000). Intellectual Imperialism: Definition, Traits, and Problems. *Southeast Asian Journal of Social Sciences* **28** (1), 23-45.

al-'Azm, S.J. (2000). Orientalism and Orientalism-in-Reverse. In: A.L. Macfie (ed.), *Orientalism. A Reader*, pp. 217-238. New York: New York University Press.

Amin, S. (2009). *Eurocentrism. Modernity, Religion, and Democracy. A Critique of Eurocentrism and Culturalism*. New York: Monthly Review Press.

Baker, J.E. (1940). Provinciality. *College English* **1** (6), 488-494.

Buruma, I. (2004). The Origins of Occidentalism. *The Chronicle Review*. http://chronicle.com/free/v50/i22/22b01001.htm, last access on 20.12.2011.

Buruma, I. and A. Margalit (2004). *Occidentalism: The West in the Eyes of Its Enemies*. New York: Penguin Press.

Clarke, J.J. (1997). *Oriental Enlightenment. The Encounter between Asian and Western Thought*. London: Routledge.

El-Haj, N.A. (2005). Edward Said and the Political Present. *American Ethnologist* **32** (4), 538-555.

Farris, S.R. (2010). An 'Ideal Type' Called 'Orientalism'. Selective Affinities between Edward Said and Max Weber. *Interventions* **12** (2), 265-284.

Galtung, J. (1971). A Structural Theory of Imperialism. *Journal of Peace Research* **8** (2), 81-117.

Galtung, J. (1967). After Camelot. In I. Horowitz (ed.), *The Rise and Fall of Project Camelot,* Cambridge, MA: The MIT Press.

Habib, I. (2005). In Defence of Orientalism: Critical Notes on Edward Said. *Social Scientist* **33** (1/2), 40-46.

Lary, D. (2006). Edward Said: Orientalism and Occidentalism. *Journal of the Canadian Historical Association* **17** (2), 3-15.

Macfie, A.L., ed. (2000). *Orientalism. A Reader*. New York: New York University Press.

Nowak, L. ([1998] 2012). The Structure of Provincial Thought. This volume, pp. 51-66.

Said, E.W. (1977). *Orientalism*. London: Penguin.

Said, E.W. (1993). *Culture and Imperialism*. New York: Vintage Books.

Smail, J.R.W. (1961). On the Possibility of an Autonomous History of Modern Southeast Asia. *Journal of Southeast Asian History* **2** (2), 73-105.

Taib, M.I.M. (2003). *On Orientalism, Culture and the Muslims: Edward Said and his Contribution to Us*. Paper presented at The Reading Group Sharing Session, 26th December 2003.

Teitelbaum, J. and M. Litvak (2006). Students, Teachers, and Edward Said: Taking Stock of Orientalism. *Middle East Review of International Affairs* **10** (1), 23-43.

Cezary Kościelniak

THE CONTEXT OF THE "THIRD MISSION"
IN THE "PERIPHERAL UNIVERSITIES"

A CASE STUDY OF THE "CROSS-BORDER UNIVERSITY"[1]

ABSTRACT. I explore the economic, social and cultural constraints of the regional mission of a university located beyond a metropolitan area or urban agglomeration, henceforth referred to as a "peripheral university." In the first part of the paper, I briefly describe the "third mission" of a university and analyze it within the context of a "peripheral university." The main constraints on the influences of regional mission and regional development are described. In the second part, I examine one type of a "peripheral university," namely a cross-border university, on a case study of a consortium of two universities: Viadrina University in Frankfurt am Oder and Collegium Polonicum – a department of Adam Mickiewicz University (CP AMU). I focus on issues like civil mission or problems of the regional contribution of a border university. I also analyze hidden-agenda concerns with respect to the trans-culture added value of the cross-border university. The ensuing analysis is based on interviews made with present and former rectors of those universities.

1. "Peripheral University" and Its Strategies

How does one measure the "peripherality" of a university? The first possible factor would be poor scientific potential, which translates into poor performance in rankings compiled by both the government (in order to determine the potential of the school) and the media (delivering their findings to would-be students). Another factor includes geographical location, where "peripheral" would mean "distant from large cities and urban agglomerations." This may, however, prove to be barking up the

---

[1] The paper was prepared with support from National Center for Science project nb. 2011/01/D/H31/01752.

In: Krzysztof Brzechczyn and Katarzyna Paprzycka (eds.), *Thinking about Provincialism in Thinking* (*Poznań Studies in the Philosophy of the Sciences and the Humanities*, vol. 100), pp. 197-215. Amsterdam/New York, NY: Rodopi, 2012.

wrong tree, since geographical location must not necessarily mean provinciality. There are plenty of examples here: University of Florida in Gainesville, Notre Dame University in South Bend or Oxbridge are all top-notch universities based in small or medium-sized urban areas where the life of the city conforms entirely with the needs of the university and students. Although rather small, such municipalities are all but "peripheral": they boast a high number of creative individuals and are relatively wealthy considering that they experience no unemployment, investment or local business development issues. Thus, "peripherality" would rather refer to the economic potential and cultural background of the school located beyond large urban areas. Precisely in this sense Zoltán Gál and Pavel Ptaček understand the term "mid-range university" (first used by Wright et al. 2008) which they apply to the redefinition of the regional mission in the cross-border areas in certain Central and Eastern European countries (Gál and Ptaček 2011). The notion of a "mid-range university" describes primarily the economic situation of the university: urban capital efficiency, research and development (R&D) potential or capacity for regional innovation. In other words, unfavorable geographic location is coupled here with poor economic and innovative potential of the school's environment:

> Mid-range universities are most often located in non-metropolitan regions or put another way, most of the universities outside the capital cities can be classified as mid-range, where the R&D potential and density of contacts are much lower and possible spillover effects emerge more sparsely. On the other hand, mid-range universities represent the keystones of regional innovation systems and are often a central element of regional innovation strategies. (Gál and Ptaček 2011, p. 1670)

We may assume, however, that it is not purely economic features that define the "peripheral" university, one has to factor in also its political and cultural background. "Peripheral" university has to deal with poor industrial potential of the area, but what is also important is that it offers neither an interesting curriculum nor attractive academic career opportunities for the scholars to develop their interests. In (2011) paper, Marek Kwiek defines an "attractive university" in terms of the diversity of obligations the institution has toward various stakeholders:

> Universities must become more attractive for ever-diversifying population of students (and respond to their ever-diversifying needs), but at the same time they need to remain an attractive place of work where one can successfully pursue her academic career. (Kwiek 2011, p. 97)

The notion of an "attractive university" calls for a wider perspective transcending the purely economic point of view. Further elaboration of the concept would define the type of the curriculum, staff-student ratio, extent of the school's international ties (cooperation with foreign institutions, international staff and students), research programs (encouraging student participation) and their position in research rankings. All these criteria would allow for determining whether the university falls under the category of an "attractive university." It follows that the a provincial university must not necessarily be classified as "peripheral" or "unattractive," although generally it is the province which may lack the environment usually provided by metropolitan areas, such as innovative technologies or access to cultural institutions. Last but not least, discourse concerning the "attractive university" must include civic activity and obligations of the university toward public and civil sphere. University is also a place where one engages in public scholarship and debate seeking to promote critical inquiry regarding local authorities and problems faced by the community. Obviously, the need for and the impact of such debate achieves a much bigger scale in large urban areas since they host a wide variety of centers for social change, think-tanks, foundations or civil associations.

One may distinguish "peripheral universities" from "leading universities," the two would be differentiated based on the following aspects: (a) regional economic and social potential as well as region location (which translates into position of the university, its influence on, e.g. spin-off companies, distance from major industrial hubs, etc.), (b) "attractiveness" of the university itself: its didactic and research offer, level of international participation and subsequent career opportunities for graduates, (c) civic mission of the university: influencing local communities through public debates, public education, pro-democratic initiatives and providing access to culture by organizing open lectures, concerts, etc. A "peripheral university" would be conceived in opposition to a "leading university," not to be confused with the phrase "central university" which is not used here due to the fact that it implies its geographical location, which is not the only feature which distinguishes such schools.

Contemporary higher education theory singles out (apart from two traditional aspects of Humboldtian mission – teaching and research) the "third mission," which emphasizes the regional aspect of university's mission. It encourages the university to respond to the economic and social needs of the region. Such needs would be satisfied mainly through academic entrepreneurship. The impact of the school on its environment is a thoroughly explored topic. Some key concepts include: Etzkowitz's *triple helix*, mode 2 knowledge production theory describing "social dispersion" of knowledge and its application to the means of production,

Burton Clark's concept of academic entrepreneurship or Richard Florida's theory of creative class and regions (see Etzkowitz and Leydersdorff 1998; Clark 1998). Comparative review of possible models of university-environment relations and academic entrepreneurship has been provided by Yusof and Jain (2010), Olechnicka and Wojnar (2008). Regional mission is not defined by the size or location of the school, it is on the agenda of both small provincial centers and universities with significant human and research potential based in large cities or inside urban agglomerations. However, its relevance is emphasized where the university plays a major role within the local economic and social system, not only as a local institution offering tertiary education but also as an employer, service provider and institution representing the region. It is precisely in less developed regions where one perceives the "third mission" as an opportunity for mutual development of the university and the region. Investment projects aiming at development of universities based in such regions have justification of their own: they spark innovative approach, thus – to quote Richard Florida – giving birth to creative class. In *The Rise of Creative Class*, Florida observed that "[. . .] investing in university infrastructure allows for much better programming of economic growth than investing in physical infrastructure" (Florida 2010, p. 229; my translation from the Polish edition)..One may say that current EU regional development policy has been founded on a dogma that, in each developing region, there must exist a university with a wide range of tools for academic entrepreneurship. Somewhat ahead of the discussion (presented in the further part of the paper), which investigates the relevance of the third mission versus two historical ones (teaching and research), one may hypothesize that some universities concede that they fall short considering their traditional tasks and embrace the "third mission" as a means for making up for their deficiencies and achieving higher status.

   Provided they do not compete for the same pool of goods, regions often join forces to enhance their economic capacity. This seems crucial for cooperation in border regions: since they belong to separate countries, competition between them is minimized but their cooperation can be put in the wider perspective of national strategy endorsing interregional and international collaboration. One example of the implementation of trans-regional policy is the cross-border university, an institution which carries out joint higher education projects exploring the potential offered by its location. Peter Arbo and Paul Benneworth (2007) explicitly describe a cross-border university as an initiative to increase the potential of less developed regions:

> However, it is clear that cross-border university activities can be at least partly successful and can help to build functional economic linkages which

in turn have the potential to help less successful regions attach themselves to more successful regions. (Arbo and Benneworth 2007, p. 76)

The potential of border regions lies also in the cultural diversity, which creates an opportunity for learning new social skills. In Western and Central Europe, cross-border universities are a frequent phenomenon. This type of cooperation can manifest itself in two ways: it is either less formal and includes joint educational programs, or it takes the form of a consortium or a federation. As for the former model, it is often applied to resolve cross-border issues. There are dozens of such initiatives across Europe, some of them include University of Innsbruck, where the law faculty offers courses in Austrian and Italian law, or the University of Bolzano in Italy, where one teaches in Italian, German and French. The purpose of those is to create opportunities for the citizens of the region inhabiting areas governed by Italian authorities and Italian citizens willing to study in Innsbruck. Formal cooperation takes the form of a "consortium" of schools thus providing a common institutional framework for collaboration. The cooperation of Viadrina University and Collegium Polonicum UAM in border towns of Frankfurt am Oder/Słubice is a sound example of such a consortium. Precisely this kind of initiative is a cross-border university in the proper sense of the phrase.

Designed to play a major role in the development of the region, universities have risen to the role of key regional stakeholders. Regional modernization policy is founded on the urge to bring the regional potential to new levels in the face of the growing supra-regional competition. However, the status of a "peripheral university" puts the institution in a rather unfavorable position within the system. In not-too-distant past, small local universities were satisfied with the prestige traditionally associated with higher education. Nowadays, in the age of economic crisis and retrenchment policy, the same schools must become more competitive and proactively cater for their future students and resources for research.[2] One may say that local politics are losing to global challenges: artificial protection of the small and local is no longer effective. Seen from the perspective of the region, the university has a tremendous cultural

---

[2] Often, the attitude toward provinciality has sentimental value, scholars sometimes provide the example of of Königsberg. There, far from the centre of the world, feeling no need leave the city, Immanuel Kant created his seminal works. While presenting the birth of modern philosophy of science, Roberto Poli (1997) shows that it was budding not only in Vienna, Cambridge or Florence, but also in Lviv and Würzburg. This implicitly indicates an historical aspect of the category of a "peripheral university," which may alter the status of the university throughout the time. Historical dynamics of "peripheral university" requires a separate historical and institutional study.

importance: having a university translates into prestige and social potential, which signifies the region's capacity to develop and prosper. Cultural dimension of the redefined regional policy (which brings new approach also to higher education issues) includes overcoming of the liberal mobility policy with its slogan "move to a better and easier life" – in the case of education and research this meant nothing else but better institution in a bigger city. The policy of liberal mobility promotes constant move between cities, companies, hubs; such dynamics may reflect the personal and institutional development pattern but less important regions are affected by an outflow of gifted individuals to the more developed environments.[3]

Another problem emerges when one compares needs of the region with those expressed by academia. Fruitful working relationship between scholars or joint research and educational programs do not automatically translate into regional development, the problem of unutilized results of trans-regional academic cooperation occurs both in highly developed countries and in the states currently undergoing transformation, e.g. in Central Europe. Good insight into the problems experienced by the former group of countries is provided in a case study of French, Swiss and German academic cooperation in ThriRhena region described by Christine Griebel (2010). The author focused on the influence of common academic sphere on, inter alia, the identity of students in Rhine region and singled out main points of mutual cooperation. Research revealed that whereas the collaboration streamlined communication and provided multiple occasions for in-depth cultural interactions between the countries of the region, it did not create a new quality one would be tempted to call a regional identity: scarce number of the surveyed students defined their identity in relation to the region. In each case, Swiss, French and German, national identity prevailed over European identity, which in turn came on top of the regional identity. For instance, 22.6% residents of Basel described themselves as Europeans, 48.8% as Swiss, 5.5% as Northwestern Swiss and only 3.7% as citizens of the region. The remaining part entered "other," such as citizens of the world (Griebel 2010, p. 19). Further, mutual exchange between the residents of the region originated primarily in effect of shopping and search for cheaper services:

---

[3] This problem was theorized by Michael Walzer in his paper "The Communitarian Critique of Liberalism" (2004). There, he embarked on a critique of liberal mobility, understood not only in its geographical dimension and related to labor migrations but also exposing its social aspect attributed to lack of inherited communitarian identity.

Students at three universities in the TriRhena region were asked about their spatial identity, cross-border social contacts and mobility. The majority of students identified themselves stronger with their national or European identity than with their status as residents of the TriRhena region. Cross-border mobility of interviewees is, however, significant, particularly concerning shopping and entertainment. Cross-border regional identity may thus be seen more in spatial activities of students than in the articulation of regional affiliation. (Griebel 2010, p. 22)

When considered as the implementation of academic integration policy within the region, TriRhena project seems far from successful. Normatively endorsed need for integration policy and common identity of academic environment within the Rhein region was basically limited to results in the field of commerce and services. This may result from the fact that wealthy cities and regions hosting "leading universities" have no cultural stimuli for tightening trans-regional bonds. This is not to say that TriRhena schools offer no intercultural programs, it just does not translate into change of cultural landscape that would manifest itself in declaration of regional identity or eagerness for closer integration. Griebel research may lead to conclusion that regional integration is not necessarily the most pressing social issue; be it on the cultural or economic level (where the existing economic mechanisms are sufficient for effective cooperation), inhabitants of the region see no need for closer ties between the regions.

Gál and Ptaček (2011), focusing on post-transformation countries, specifically on Czech-Hungarian borderland, pointed at other problems experienced by cross-border universities. They argue that the sole capability to establish such relations in provincial areas is much more reduced than in the capital city regions, mainly due to the relatively low economic and R&D potential:

Nevertheless, in the CEE countries with centrally coordinated innovation policies, non-capital city regions have substantially different innovation and third role histories and policy options than those of capital city regions. (Gál and Ptaček 2011, p. 1674)

The authors examined the cooperation between universities based in Hungarian (Southern Transdanubia) and Czech (Central Moravia) borderland. They have demonstrated that the increase of R&D potential had no direct impact on regional development on any of the sides of the border. In the regions with lesser "techno-economic maturity," output of academic research programs cannot be absorbed by the regional economy (Gál and Ptaček 2011, p. 1686). Thus, for the reasons attributed to the low adaptation ability of the regions' economy and regardless of the effort on the part of the university, no linkage has been created between innovative research and local economy. Strategy adopted in this respect by "peripheral universities" contributes to the viscous circle: lack of mature economic environment causes incapability to overcome the peripherality of the local schools which in turn are not able to produce the added value that could be consumed by the local economy. The two above examples of Western and Central Europe show the difficulty one experiences in the process of creating bonds between universities and regions in both highly developed and post-transformation countries. This difficulty may be attributed to the weakness of the region itself, on the one hand, but it is related to the discrepancy of expectations expressed by the university and the needs of the region, on the other.

One chance for successful joint region-university undertakings is coordination of regional and higher education policies where strength of the school, together with its educational potential, is coupled with the growth factors of region attractiveness (see Kościelniak 2011).

To elucidate the topic, it may be of use to bring forward Polish experiences in this respect. In the past 14 years, Poland has followed the policy of setting up new regional vocational higher education institutions (in Polish: *Państwowe Wyższe Szkoły Zawodowe*) with the purpose of educating professionals whose expertise would later contribute to the development of the regions. Some schools got to specialize in training experts of certain fields, e.g. aviation technology or IT, which effectuated in smooth transition of graduates to professional career and further education. This strategy seemed to be successful until many local authorities, driven by the ambition to establish their own local university, launched lobbying campaigns at the ministry which resulted in uncontrolled growth of the schools thus depriving them of their core functions. Examples of the process include establishment of two such schools in Tarnobrzeg and Sandomierz, one located just a dozen kilometers from the other. Another fitting example is the State School of Higher Education in Oświęcim based in a close distance to Cracow and Silesian District, both being academic hubs of southern Poland (e.g. departments of University of Silesia, Silesian University of Technology and University of Economics in Katowice are spread out in various cities in the Silesia District and produce a network effect). Similar situation persists in Gniezno, where local State School of Higher Education co-exists with Collegium Europeum, a branch of Adam Mickiewicz University. What is more, Gniezno is located in the proximity of Poznań, just forty kilometers away. In effect, two state universities compete for the students, in the times of looming demographic slump it constitutes a serious problem. The produced facts provoke questions concerning rationale behind such policy. Is it reasonable to support two public universities in a small town having fifty thousand residents? Are the times of demographic low and crisis, which is likely to last over a decade, a well-timed moment for supporting both academic centers, or should they be rather merged into one organism?

A fine example of the matter at hand is provided by German Leuphana University located in Lüneburg, a town around 70 thousand residents in Lower Saxony. In the 90s, Lüneburg hosted two schools, a university and a polytechnic, who were, however, no match to a not-too-distant Hamburg. The authorities decided to merge them, thus establishing Leuphana University specializing in research within the field of sustainable development (for a related issue, see Adomβent 2011). Although opposed by many, the idea came as a rescue: if not for the merger, one may safely assert that both institutions would have perished. The case of Leuphana clearly indicates that setting up several frail institutions will never cure the peripherality of the universities; it is exactly the opposite: merging and redefining the management strategy may eventually bring success.

Rapid changes on the higher education market force the State Schools of Higher Education to compete with the universities belonging to traditional higher education system. In effect of demographic low, competitiveness is predominantly determined by the choice of school made by the future student. Therefore, such schools cannot withstand competition with larger academic centers which in turn strictly coincides with collapse of the regional mission they have been designed to fulfill, namely training experts contributing to the development of the region.

2. Regional Mission of Higher Education and Its Constraints

Throughout the last twenty years, regional policy of OECD countries has seen a radical shift of dynamics. It is not only of declarative value, the report published by OECD *Regions and Innovation Policy* (2011) reveals that in European Union countries, the change has brought about a major increase in spending: from 4% in the 90s to 25% as of today, with the growth predicted to continue:

> In EU countries, the availability of Structural Funds has helped regions mobilise their assets for knowledge-based growth. Innovation has become one of the major pillars of EU regional development policy. From 1989-1993, approximately 4% of regional policy funds were devoted to innovation (2 out of 50 billion). The share of broadly defined innovation-related spending for the period 2007-2013 is projected to be approximately 25%, totalling around EUR 86 billion. (OECD 2011, p. 32)

Political shift and its financial implications caused the need to redefine the mission of the university: it is now becoming a partner in economic activity undertaken by regions. In their book *Understanding the Regional Contribution of Higher Education Institutions* (2007), Arbo and Benneworth picture universities as inspiring actors actively participating in modernization within the field of regional knowledge, industry and innovation. These various regional roles manifest in the following tasks (Arbo and Benneworth 2007, p. 75): "representing own interests," meaning introducing changes which open the possibility to absorb regional tasks by the university itself; further, "selling governance support services," i.e. offering consulting and analyses enhancing the region management; "managing own business," understood as university entering into economic relations with external stakeholders; and "supporting communities of 'good citizens'," which authors conceive as personal involvement of highly skilled academic staff in the economic projects thus creating a network of mutual dependence. They are not only concerned with economic perspective but also underscore the role of the university as the major agent within the democratic system:

> The university learning experience is the promotion of meritocratic ideals and the critical thinking skills necessary for a well-functioning democracy [. . .], creating critical knowledge of [the] communities but also representing those communities to the wider public sphere. (Arbo, Benneworth 2007, p. 82)

Lastly, cultural mission of the school is related to sustainability agenda, i.e. education and research seeking to promote the idea of sustainable development: "Finally, universities have involved themselves with institutions and activities directed toward local and regional sustainability" (p. 83). Regional mission breaks down into activity carried out within the areas of economy, civil society (related to support democratic process) and culture (with emphasis on sustainable development issues).

Chatterton and Goddard (2003) demonstrate that such academic tasks as teaching, research and service to the community should produce new values for the empowerment of the regions: skills, innovation and culture of the community. The "third mission" should be therefore translatable into economic growth. Such is the meaning attributed to the third mission in the already cited study of Gál and Ptaček:

> The third mission of universities in relation to sub-national economies and societies has been widely justified in terms of the development of the knowledge economy and the significance of the regions in economic development. The third mission calls for universities to transform

themselves into economic institutions by taking on specific tasks such as greater technology transfer, more patenting, visible employment and commercial outputs [. . .] (Gál and Ptaček 2011, p. 1673)

The above characterization presents a bright future of regional mission and therefore requires thorough critical analysis. Undeniably, the "third mission" is not a universal cure and encounters constraints, primarily related to the ongoing crisis. Marek Kwiek observed that in the age of prevailing retrenchment policy, regions' ability to cope on their own will be increasingly tested in the future, especially considering limits imposed on aid money:

European model of regional development undergoes a serious reconstruction: presently, regions are encouraged to take care of their own issues and can no longer count on traditional, compensational or balancing, roles (and funds) of the state. (Kwiek 2010, p. 272)

If the regions are on their own, so are the regional universities. Further, Kwiek examines the status of the "third mission," which in the end of the day boils down to the support of the first and the second mission.

Is what we witness a new role of the university or maybe just a new way to fund its two traditional and basic tasks? If it is really a new role, [. . .] such activity would be fully entitled to receive state funding as it is the case with both traditional missions, namely teaching and research. (Kwiek 2010, p. 274)

Here, Kwiek exposes the first doubt related to the matter at hand: is the "third mission" a genuine academic "mission" or just an implication of entrepreneurial strategy seeking additional funding for one's activity? Assuming that it is the sole responsibility of the universities to meet the requirements imposed by the third mission, the key issue concerns its resources. If one examines approach to this problem in Britain, where the crisis forced the authorities to scrap funding of multiple non-market higher education institutions (including open universities committed primarily to public education), one may assert that the "third mission" is quite unreliable foundation of academia because, in the face of hardships, the state will be reluctant to secure the secondary activity of the school. Marek Kwiek's argument may be construed as follows: the state's acknowledgement of the university's mission translates into higher education funding policy where some expenditures meet "boundary conditions" set in the budget.

Another constraint distinguished by Gál and Ptaček concerns academic entrepreneurship, which in the case of a "peripheral university" is a highly challenging task considering the immature environment and weak economic background:

This means that in most of the non-metropolitan CEE regions the regional innovation systems are still weak similarly to the university and industry linkages. [. . .] The increasing R&D investment does not have significant and immediate impact on local growth if the appropriate industrial structure, on which the university-industry links can be built, does not exist in the region. Another main constraint is that the financial endowments of mid-range universities in CEE are weaker than their western counterparts and their impact is also highly dependent on their region's absorption capacity, which is related to their institutional set-up, techno-economic characteristics and economic specialization. (Gál and Ptaček 2011, p. 1676).

The authors of the study bring attention to the following constraint: despite of creative potential and relatively abundant funding of academia, results will have little impact on the region. This argument implicitly undermines Florida's reasoning who believes that creative environments directly translate into growth of regional innovation.[4] This is difficulty of the transformation countries, where underinvested industry cannot be contributes to the fact the effort put in building academic entrepreneurialism.

Putting aside economic and infrastructural constraints, "peripheral universities" experience also certain cultural problems. First of those would be *unsynchronized "regional mission" and higher education policies* EU countries. Above, we hinted on discrepant needs of the regions and expectations of universities. Regions require support which would help overcome such local problems as unemployment and scarce additional funding. In turn, universities – to employ the phrase coined by Kwiek – are set on enhancing their attractiveness, both in the field of teaching and research. In reality, this means: pursuing the idea of perfection (conceived as meeting research standards acclaimed by international scholar community), acquiring outstanding scientists, and recruitment of would-be scholars from among gifted students. These goals can only be achieved after completing long-term projects involving multi-generational teams of scholars whose work spans through years or even decades. Meanwhile, regional policy is determined by budget years, duration of financial projects or election cycles. Regions adopt time frame which is undeniably narrower than the temporal perspective of academia. Quick results, so desired by politicians (especially with regard to reduction of

---

[4] Critique of Florida was offered by Peck (2010). In his text, he shows that changes encouraged by Florida are compatible with neoliberal policy, e.g. experts working in creative professions will move to smaller communities for better salary; hence, neoliberal mechanisms are here unavoidably present.

unemployment), may not always be achieved even by the most entrepreneurial universities of all. Further, higher education policy a model of academic career where production of scientific results is the main criterion of success, teaching comes as far second and focusing on regional activity is barely noticed. The "third mission" is not equal to teaching and research: it does not translate into academic career. A involved in the spin-off company will not be rewarded by promotion; nor will he join the international community of prominent scientists because regional research targets local communities. One may encourage this sort of activity and appeal to possible benefits of the parties involved, but such activity will not necessarily directly boost one's academic career or march up the ladder of academic hierarchy. Marek Kwiek (2010) supplies a relevant comment in the footnote of his book:

> There is a mental barrier in educational institutions which may be difficult to break, it deems regional involvement somewhat less important [. . .]. Regional dimension of research and teaching should finally cease to be labelled as less prestigious. (Kwiek 2010, p. 273)

Reasons for lesser concern for region-related research may be attributed not only to unsynchronized policy for the development of academic career but also to the benefits related to participation in international scientific teams. This negative trend could be reversed by paradigm shift within higher education policy which would include redefinition of university's mission and provide enduring legal, financial and institutional foundations for academic involvement in the "third mission." The problem is, however, that conducting sponsored research which fits in one's career scheme requires a difficult and extremely rare compromise, interests of scholars and business are not easily reconciled.

Another cultural constraint experienced by "peripheral universities" is lack of permanent and consolidated academic environment. This problem has two separate causes, one has legal, the other academic background. Scholar is allowed to be employed only in two institutions, usually one takes another job at the "peripheral university" thus failing to invest there his entire potential. There is also a problem with cities hosting universities: permanent presence of university professors lends academic climate to the city, allows the city to develop its social and urban character. Unfortunately, scholars are but rarely based in their places of employment and cannot contribute to the less official dimension of academia which is a life of its own, lived in cafes and clubs, bringing unique cultural vibe to the place.[5] Commuting professors are an especially frequent phenomenon if one considers peripheral and cross-border universities, the case at hand is no exception here: in Frankfurt (Oder) and Słubice, most professors (at Viadrina this includes also the president of the school) commute from larger academic centers. Not permanent residents of the town, professors neither leave their mark on the place nor create an intellectual milieu or cultural tissue, e.g. public debates are never organized since no one is eager to work after hours and devote their time to the community. Such municipalities resort to offering scholars apartments for lower rent but this incentive proves to be insufficient. One may consider this issue a minor one, but it pictures a dissonance between the declared formal affiliation with the given academic centre (by indication of the place of employment) and actual evaluation thereof (or at least of the environment one has come to work in). Let us emphasize that the staff of such aforementioned leading universities, as Gainesville and Ann Arbor in the U.S. or oft recalled Oxford and Cambridge is basically based in its place of employment or in its close proximity.

Finally, cross-border universities face one important institutional and cultural problem: *incoherence of national higher education systems*. In a way, Bologna Process offered certain solutions of the issue by allowing for unification of education standards: amendment to the Polish Higher Education Act permits to organize courses concluded by awarding a common diploma. Nevertheless, problems related to academic culture remain. Interviewed staff of Viadrina and CP AMU indicated that German side was at times unwilling to accept English as the language of instruction, which quite understandable considering the fact that European University Viadrina, regardless of its official name, remains an institution affiliated by Federal Government of Brandenburg. One should also look into the issue of different academic cultures. Heidi Fichter-Wolf (2007) concludes that common informal rules of academic culture should be subsequently institutionalized and come to be reflected on the formal level.[6] (Fichter-Wolf 2011).

---

[5] This topic is explored by Richard Florida who shows that the city seeking to elevate its status must cater for art world celebrities who will lend new dynamics to its districts. Richard Florida is by no means original, back in the communist times in Poland one engaged in similar activity. For instance, the authorities of Szczecin invited the eminent writer Jerzy Andrzejewski to settle in the city and offered him excellent material incentives. Andrzejewski, however, returned to Warsaw after failing to find in Szczecin inspiration for his spiritual development.

[6] H. Fichter-Wolf, "Europeanization from Bottom Up: Learning from Everyday Experience in Border-Unis: Intercultural Learning and Institution Building," presented at BorderUni

3. Cross-Border University as a Type of "Peripheral University." A Case Study of European University Viadrina and Collegium Polonicum

This part of the paper examines a consortium of two cross-border universities: Collegium Polonicum at Adam Mickiewicz University (CP AMU) and Viadrina. It presents interviews with the present and former rectors of these universities (related issue, see: Bielawska and Wojciechowski 2007). The interviews were conducted in 2011 and 2012. Their aim was to look behind the official agenda of the border university (provided in official statements, documents, statistics), to try to depict the real condition of these institutions and to record the process of its institutional and cultural change from the beginning of the 1990s till 2012. The following matters were discussed: the perspectives of consortiums of cross-border universities, their problems, funding issues, and Polish-German cooperation.

At the outset, let us provide the background of institutional changes experienced by both schools. Collegium Polonicum was established in 1996 as an institution complementary to European University Viadrina. Consortium was run jointly by University of Wrocław and Adam Mickiewicz University until the latter took exclusive responsibility for the enterprise. It has quickly won acclaim of Polish students. Before Poland's accession to the European Union, when, contrary to present day, exchange programs were a rarity, studies in Słubice and at Viadrina offered unique, bilingual curriculum and infrastructure (lecture halls, library, dorms) that by Polish standards was second to none. Even at the turn of the century the schools' graduates were guaranteed a quick access to welfare society. Lawyer or Germanist with comprehensive knowledge of the language and culture experienced no troubles finding a job. Moreover, Polish students were free to cross the border and participate in Western academic . Joint cooperation abounded in research and publishing projects (Logos, a Berlin-based publishing house, offers publications of joint Polish-German research). What is important, however, is that both institutions are pro-active in the process of European integration, Viadrina and Collegium Polonicum has hosted numerous political figures including presidents of both Poland and Germany. Słubice and Frankfurt (Oder) has come to epitomize the politics of reconciliation and integration. One of the

---

Workshop, December 8, 2011, Viadrina, Collegium Polinicum, Adam Mickiewicz University, Poznań, Poland.

interviewed rectors stated that Słubice, once nothing but a sordid marketplace, turned into a place vital to politics of both countries. This came as a result of the fact that local academic culture is embedded in the broader process of Europeanization. It is not difficult to notice that Viadrina/CP AMU achieved a unique status in Central Europe. One did not succeed to apply similar institutional model elsewhere, although in Polish-German borderland there exists cooperation between State School of Higher in Nysa and Lausitz University, another example of such endeavor includes University of Białystok opening its branch in Lithuanian Vilnius. But despite of the efforts made by Polish and Austrian Ministries of Science, the idea to open East European university, conceived as a common platform for the lands of former Galicia (Poland, Slovakia, Ukraine and Austria), has never been materialized. One of the rectors opined that it shows that success of Viadrina/CP AMU was in itself not a natural process, but rather involved putting into life a political idea which, combined with political and financial support, allowed the consortium to gain political meaning. Interviewee insisted, however, that the price for this was high. Lack of political will, but also awareness of the ensuing costs, suppressed the similar project on the eastern border. After Poland has joined the European Union, the idea of Europeanization lost momentum, both in its academic and political aspect; finally, entering Schengen Area sealed its fate. Freedom of travel and access to Socrates programs, regional programs and large framework programs contributed to the fact that borderland ceased to be of major importance and had to confront the issue of its peripherality. First, such schools are based in regions with social and economic dysfunctions, high unemployment rate and weak economic environment. Both Frankfurt (Oder) and Słubice record high migration rate, one of the interlocutors termed the outcome as the "pensioner towns." Another interviewee quoted statistics saying that almost half of the youth finishes high school elsewhere, one presumes that in a larger city. However, the most painful issue is related to the fact that in the wake of Poland's accession to the EU these universities have ceased to be attractive. To add insult to injury, the school has no department of exact sciences which today is a major setback.

It is thought-provoking that neither of the interviewees has agreed with the idea that Viadrina/CP is in a sense competing with Poznań headquarters of AMU. One of the rectors argued that in this respect both universities compete with Berlin which is a popular academic hub attracting large numbers of students. Obviously, it cannot be denied that CP AMU is in crisis, the majority of Collegium Polonicum students is recruited from among inhabitants of Słubice and its surroundings, this indicates that the schools has only rather regional impact. The question of competitiveness in times of demographic low is not illegitimate, one of the German rectors conceded that Polish and East European students may come to the rescue of the consortium. This may be so, but at the same time the same group of potential students can be targeted by both CP and AMU in Poznań: those may cooperate, but they remain separate institutions. Here, we touch upon a confusing matter, namely cooperation of both schools. Whereas CP and Viadrina run joint programs, CP has programs on its own and invites German lecturers to participate in them. Fields of competition between those entities remain largely unclear.

One of the rectors stated that a reasonable strategy for the future would include, firstly, establishing a department of exact sciences focusing on environmental research, and secondly, closer cooperation with Viadrina aiming at further integration and establishment of one joint university in the future. As for scientific development, he trusts that the existing grant & project system is quite efficient. In the future, the submitted projects would constitute research potential, which, materialized, could eventually help to drop the present peripheral status of the schools. Funding would first be provided by AMU and Viadrina. After it grows stronger, the institution would generate its own research resources. Simultaneously, schools would become involved in regional mission and cooperate with such centers located along the Odra river as Szczecin or Geiswald. Bearing in mind information provided in the first part of this paper pertaining to peripheral universities in the Czech Republic and Hungary, implementation of such strategy may encounter problems: weak economic and cultural background, non-existing academic culture, incompatible higher education systems and cultures. Another interlocutor pointed at the need for new political approach which would once more turn them into institutions of political significance. In his opinion, singular character of Viadrina/CP AMU distinguishes it from other similar entities, but for the time being this uniqueness has been lost. Completely redesigned mission of the university would focus on the education concerning border-related issues: border protection, border security or cross-border criminology. One of the interviewees brought to the light political marginalization of cross-border university. As of today, neither Poland nor Germany discuss new roles and missions of such institutions. On the one hand we should not lost the perspective of the special role of the Collegium Polonicum/Viadrina, which concern political issues, e.g. Polish – German reconciliation and creating new trans-cultural curricula. On the other hand, we should notice the fact that cross-border universities have already fulfilled their mission, united Europe of today is a place where the choice of the university is only limited by the student's level of affluence. Lesser concern for the idea of cross-border university may be also attributed to the deepening crisis and possible return of the institution of state to its preponderant position. Obviously, there is little cultural need for such initiative. Comparison of current condition of Viadrina/CP with Christine Griebel's findings reveals that what we face is not a negligence, but the usual course of events where exchange has taken the shape of commercial activity and provision of services; call for closer regional or national integration would not chime with the expectations of the residents involved, as it was the case with the Swiss-German-French borderland.

Conclusions

The issue of "peripheral universities" is closely related to the idea of university's "third mission" as well as problems related to the academic culture. Consequently, it is also a problem experienced by cross-border universities. Current higher education policy champions competitiveness (which in Poland is reflected by distribution of research resources by the means of grant competition); this affects both leading and peripheral universities and forces these institutions to devise strategies giving them new possibilities for development. It seems, however, that cross-border universities are in dire straits for a number of reasons. (1) They have to redefine their missions and identity in the prevailing political landscape within European Union. (2) If such universities were to join the system of academic competitiveness (in the field of both student recruitment and "third mission" activity), they will experience much more troubles trying to meet those criteria than the non-peripheral universities, one reason for this is the aforementioned weakness of academic culture. (3) Economical and social immaturity may be a serious encumbrance for implementation of academic entrepreneurship in the regions featuring "peripheral universities," this concerns also Viadrina and CP AMU. This difficulty is in a sense independent, overcoming it does not lie within the powers of the university.

Adam Mickiewicz University
Institute of Culture Studies
ul. Szamarzewskiego 89a
60-568 Poznań
Poland
e-mail: cezkos@amu.edu.pl

REFERENCES

Adomβent, M. (2011). In Search of the Knowledge Triangle for Regional Sustainable Development: The Role of the Universities. In: A. Barton and J. Dlouha (eds.), *Multi-Actor Learning for Sustainable Regional Development in Europe: A Handbook of Good Practice*, pp. 5-15. Guilford: Grosvenor House Publishing.

Arbo, P. and P. Benneworth (2007). *Understanding the Regional Contribution of Higher Education Institutions.* Paris: OECD/IMHE.

Bielawska, A. and K. Wojciechowski, eds. (2007). *Trans-Uni.* Berlin: Logos Verlag.

Chatterton, P. and J. Goddard (2003). The Response of the Universities to Regional Needs. In: F. Boekema, E. Kuipers and R. Rutten (eds.), *Economic Geography of Higher Education: Knowledge, Infrastructure and Learning Regions*, pp. 19-42. London: Routledge.

Clark, B. (1998). *Creating Entrepreneurial Universities. Organizational Pathways of Transformation.* New York: Pergamon Press.

Etzkowitz, H. and L. Leydersdorff (1998). The Endless Transition: A "Triple Helix" of University-Industry-Government Relation. *Minerva* **36**, 203-208.

Florida, R. (2010). *Narodziny klasy kreatywnej.* Warszawa: NCK

Gál Z. and P. Ptaček (2011). The Role of Mid-Range Universities in Knowledge Transfer in Non-Metropolitan Regions in Central Eastern Europe. *European Planning Studies* **19** (9), 1669-1690.

Griebel, Ch. (2010). Grenzüberschreitende regionale Identität in der Regio TriRhena von Studierenden der Universitäten Basel, Freiburg im Breisgau und Mulhouse. *Geographica Helvetica* **65**, 15-23.

Fichter-Wolf, H. (2007). Od przestrzeni przygranicznej do przestrzeni wiedzy. Transgraniczna współpraca szkół wyższych jako przykłady regionalnych centrów kompetencji w Europie. In: Bielawska and Wojciechowski (2007), pp. 58-76.

Kościelniak, C. (2011). Transformacje regionalnej dynamiki wyższego szkolnictwa artystycznego: Przypadek Wielkopolski. *Człowiek i społeczeństwo* **32**, pp. 59-71. Poznań: Wydawnictwo UAM.

Kwiek, M. (2010). *Transformacje uniwersytetu. Zmiany instytucjonalne i ewolucje polityki edukacyjnej w Europie.* Poznań: Wydawnictwo UAM.

Kwiek, M. (2011). Co to znaczy atrakcyjny uniwersytet? In: C. Kościelniak and J. Makowski (eds.), *Wolność, równość, uniwersytet*, pp. 73-110. Warszawa: Instytut Obywatelski.

OECD (2007). *Poland.* Review of Tertiary Education.

OECD (2011). *Regions and Innovation Policy.* OECD Reviews of Regional Innovation.

Olechnicka, A. and K. Wojnar (2008). Rola uczelni wyższych w regionie. Doświadczenia międzynarodowe. In: T. Markowski (ed.), *Rola uczelni wyższych w przestrzeni miast*, pp. 14-33. Warszawa: KPZK PAN.

Jain, K. and J. Yusof (2010). Categories of University-Level Entrepreneurship: A Literature Survey. *Institutional Entrepreneurial Managing* **6**, 81-96.

Peck, J. (2010). Zastrzyk kreatywności [The Creativity Fix]. In: *Ekonomia kultury*, pp. 99-123. Warszawa: Wydawnictwo Krytyki Politycznej.

Poli, R., ed. (1997). In itinere. *European Cities and the Birth of Modern Scientific Philosophy. Poznań Studies in the Philosophy of the Science and the Humanities*, vol. 54. Amsterdam: Rodopi.

Wright, M., B. Clarysse, A. Lockett and M Knockaert (2008). Mid-Range Universities' Linkages with Industry: Knowledge Types and the Role of Intermediaries. *Research Policy* **37** (8), 1205-1223.

Krzysztof Brzechczyn

ON COURAGE OF ACTIONS AND COWARDICE OF THINKING

LESZEK NOWAK ON THE PROVINCIALISM OF THE POLITICAL THOUGHT OF SOLIDARNOŚĆ

ABSTRACT. In the opinion of many Western observers (e.g. Timothy Garton Ash) as well as Polish authors (e.g. Zdzisław Krasnodębski), the political thought of Solidarność was a mixture of ideas taken from different ideological traditions (right and left). What, in the aforementioned authors' opinion, was a reason for pride was an object of criticism by Leszek Nowak, the eminent Polish philosopher, engaged in the Solidarność movement. One of his most important charges against the political thought of this movement was its intellectual provincialism and its inability to propose something new and fresh. The purpose of this paper is to present Nowak's reflection on the political thought of Solidarność in years 1980-1981. I show that he presses three general kinds of objections. According to Nowak, the political thought of the movement had formal-internal deficiencies (it provided no clear theoretical vision), cognitive deficiencies (it was incapable of offering a diagnosis of the situation) and policy deficiencies (it was incapable of indicating the appropriate course of action).

1. Introduction

The article is an attempt at an interpretation of Leszek Nowak's criticism of the political thought of the Solidarność movement in Poland in the years 1980-1981. Provincialism in political thinking was one of the most important charges made by Nowak against the political program developed by the Independent Self-governing Trade Union "Solidarity" (NSZZ Solidarność) and also against the wider political thought of the opposition in Poland between 1980 and 1989. The article discusses Leszek Nowak's criticism of the social ideas of Solidarność with regard to its formal (internal), cognitive and policy dimensions.

According to Hannah Arendt, a defining trait of modern revolutions is the conviction that a revolution as a historical event triggers a new

In: Krzysztof Brzechczyn and Katarzyna Paprzycka (eds.), *Thinking about Provincialism in Thinking* (*Poznań Studies in the Philosophy of the Sciences and the Humanities*, vol. 100), pp. 217-234. Amsterdam/New York, NY: Rodopi, 2012.

experience unknown before. In her analysis of the American and the French Revolution the novelty was the experience of freedom:

> This relatively new experience [. . .] was at the same time the experience of man's faculty to begin something new. These two things together – a new experience which revealed man's capacity for novelty – are at the root of the enormous pathos which we find in both the American and the French Revolution, this ever repeated insistence that nothing comparable in grandeur and significance had ever happened in the whole recorded history of mankind. (Arendt 1963, p. 27)

Let us systematize Arendt's remarks on the revolutions by indicating three levels of novelty of a historical event:

- *material level* is reached when a new, hitherto unknown, type of social praxis emerges, or when some old social praxis formerly forbidden by an oppressive social system re-emerges;
- *subjective-psychological level* consists of private opinions, convictions, emotions and feelings of individuals taking part in the historical event that constitutes a fundamental (generational, national, etc.) experience of their lives. This experience can serve the role of a founding myth, which legitimizes the social order initiated by the historical event – in this case by a revolution;
- *ideological-institutional level* in the form of an ideological manifesto, revolutionary ideology or political platform solidifies the two previous levels of novelty and becomes the basis for new institutions posed to change the existing social order.

The concurrent attainment of these three levels results in a fundamental and durable social transformation. A question can be asked whether these three dimensions of novelty can be found in the Solidarność revolution in Poland in the years 1980-81, which was surely a unique phenomenon in the history of real socialism. Timothy Garton Ash, one of the eye-witnesses of historical events taking place in Poland at that time, noted that

> What happened in Poland does not fit into any of these pre-formed western moulds and [. . .] rather than manipulating the Polish revolution until it fits into our existing categories, we might do better to adjust our categories until they fit the Polish revolution. (Ash 1983, p. 307)

The proof of the unique character of the Solidarność movement can be the multitude of terms used to describe it: 'workers' protest', 'workers' revolution', 'national liberation uprising', 'social movement', 'civil society', 'ethical and religious community', etc. (Ciżewska 2010, pp. 23-68; Dudek 2002). This state of the affairs was recognized in the popular awareness of participants in the Solidarity revolution. For them, as attested in numerous memoirs and diaries, the participation in the strikes was one of turning points of their lives, which made them subscribe to one of the sides of the conflict and re-establish severed national and social ties. This experience was perfectly reflected in the word 'Solidarity', which signifies sharing of responsibility for the fates of others, rebuilding grass roots interpersonal cooperation and straightening out the public language – all of which had been formerly impossible in a society ruled by the communists.

Were the novelty of the Solidarność social praxis and self-awareness of its participants verbalized in any form of an equally novel and coherent ideological project solidified in new social institutions after 1989? Ash in his description of the Solidarność political ideas observes that:

> The Poles in fact produced a quite original mixture of ideas drawn from diverse traditions. In politics, they dove to the central principles of liberal democracy, but they combined this with proposals for a kind or radical devolution, social control and local self-government [. . .]. For culture and education, their ideals could best be characterised as conservative-restorationist. In economics, they wished to combine the market, self-government and planning. (Ash 1983, p. 338)

In a similar way the Solidarność phenomenon is perceived by Zdzisław Krasnodębski, for whom the ideas of Solidarność were a Polish embodiment of republicanism combined with the notion of participatory democracy. Republicanism, in this perspective, was founded on the same value as liberalism, i.e. freedom of individual citizens. Krasnodębski, however, defines this guarantee of freedom in a different way:

> In the republican tradition freedom is not understood as the lack of external interference, but as independence of a foreign power which constraints one's freedom, even without interfering with this freedom. The very possibility of autocratic coercion, and not its actual use, is considered to be a state of non-freedom, i.e. subjugation. (Krasnodębski 2003, p. 281)

An individual's freedom is guaranteed only by the freedom of all citizens, i.e. independence of a nation consisting of individuals. This is why the state is not perceived as a threat to individual civil liberties or as a neutral framework permitting the fulfillment of individuals' selfish interests, but as the common good – *res publica* – of all citizens.

The concept of participatory democracy, developed by the so-called New Left movement, is regarded as a complement to the model of liberal democracy, helpful in overcoming the deficiencies of the latter: alienation of the elites, growing bureaucracy and withdrawal of citizens from public life (Krasnodębski 2003, pp. 74-76). Freedom in the years 1980-81 was understood as a possibility of self-management, independence of arbitrary

communist rule and joint action aimed at realization of collective aims. In
this sense, the political thought of Solidarność was a combination of a
right-wing idea (republicanism) and a left-wing idea (participatory
democracy). According to Krasnodębski:

> Solidarność [. . .] was [. . .] a liberation-oriented republican movement
> with a range of unique features. It shared with liberalism, the idea of
> freedom of the individual [. . .], however it understood that freedom
> differently, knowing that one cannot be free as an individual, as long as
> Poles as citizens will be dependent on the communist authority.
> (Krasnodębski 2003, p. 293)

Thus, considering the conclusions of both aforementioned authors, the
Poles did not invent anything new. The intellectual novelty of the
Solidarność revolution lies in the very combination of ideas conceived
somewhere else, i.e. in the intellectually developed West, where these
ideas usually contend against each other. The ideological novelty of the
Solidarność revolution is of secondary nature. It is a "second-hand"
novelty comprising ideas already existing on the intellectual market.

2. Why Was the Social Thought of Solidarność Provincial?

What, in the aforementioned authors' opinion, was a reason for pride was
an object of criticism in Leszek Nowak's view. On the one hand, Nowak's
analyses were based on his construct of the idealizational theory of science
(Nowak 1980b; Nowak and Nowakowa 2000), and on the other hand, on
the theory of historical process known as non-Marxian historical
materialism (Nowak 1983c; 1991).

The idealizational approach to science involves a conviction that a
scientific theory is neither a generalization of facts nor a hypothetical
deductive system but that it begins with a radical distortion of the reality.
In the most idealized model, an assertion is made how a studied
phenomenon depends only on its principal factor. This assertion is a
conditional statement. Its antecedent contains counterfactual assumptions
that the impact of any secondary factors on the studied phenomenon is
ignored. The consequent of this statement shows how the examined
phenomenon depends on the principal factor. The resulting idealizational
statement holds only under the accepted idealizing assumptions. This
simple picture of the phenomenon as dependent only on the principal
factor is only gradually modified. Idealizing assumptions excluding the
influence of the secondary factors are waved, and it is shown how factors
that were ignored in the initial statement modified basic dependencies. In

this way, a scientific theory is developed, which consists of a sequence of statements that approach the studied phenomenon with greater accuracy.

This approach to science preferred a certain type of meta-scientific attitude described by Leszek Nowak in his columns (1973a-f; 1974a-c). What is especially crucial for our considerations is the fact that the basic component of this attitude was the prohibition of eclecticism combined with the directive of tolerance, which should not to be confused with intellectual chaos. According to the idealizational theory of science, a given phenomenon can be conceptualized from various points of view. These points of view are generated by philosophical orientations that determine the type of main factors that are significant for the given phenomenon and the order of importance of secondary factors. Pluralism of philosophical perspectives is important since the principal factor in one philosophical orientation can lead to the modification of a theory assuming another philosophical perspective. Theoretical and ideological pluralism must not be confused with intellectual chaos. The modification of a theory due to a factor from a different philosophical perspective is not an eclectic inclusion of this factor into the class of factors recognized in a researcher's philosophical perspective as principal ones. Rather, it is a concretization of the initial model of the theory. This modification consists in revealing silently accepted idealizing assumptions, removing them and disclosing the ways in which the principal factor in a different philosophical perspective modifies the impact of the factor recognized as principal in the researcher's philosophical perspective. In other words, the principal factors in different philosophical perspectives are the sources of modification (as secondary factors) of the theory developed by a researcher maintaining his own philosophical perspective.

Three important characteristics of such a scientific approach can be distinguished: courage, tolerance and capacity for modification. Courage is necessary to accept the idealizing assumptions, which "parenthesize" practical knowledge about the effects of secondary factors on a given phenomenon and reveal the impact of the principal factor. Tolerance grants others the right to develop alternative philosophical perspectives. Finally, capacity for modification consists of incorporating principal factors from other philosophical orientations into the theoretical set of one's own assertions, where they are regarded as secondary factors.

These characteristics of intellectual activity were considered by Leszek Nowak to be universal or at least applicable in philosophy and political thought. However, apart from the *formal-internal criteria* mentioned earlier, philosophy and political thought should also fulfill *cognitive criteria*, i.e. accurately recognize a given social state, and *policy criteria*, i.e. propagating the vision of a social order, institutional and legal systems pursuant to the recognized social state. Now let us interpret Leszek Nowak's reflection on the political thought of Solidarność using the aforementioned criteria.

2.1. *On the Formal-Internal Deficiencies of the Political Thought in the Solidarność Movement*

In his article under the meaningful title "The Price of the Lack of Perspectives," Leszek Nowak criticized the project of the NSZZ Solidarność platform:

> There is no clear social vision. For a year, we have been witnessing the growth of the most formidable social movement in the history of socialism, bringing hope to hundreds of millions of people living permanently in the darkness of enslavement. Yet this movement has not measured up intellectually and ideally to itself. This is clearly visible in the "Kierunki" (Courses of Action) document of the Solidarność movement, which in the beginning states that: "The best national traditions, ethical principles of Christianity, political calls for democracy and socialist thought are the four main sources of our inspiration." If only it were an inspiration for a qualitatively new idea! After all, the Solidarność movement was an entirely novel historical phenomenon. The movement, however, starts and ends with the four mentioned types of ideas. They are not the sources of inspiration for the movement but its results. (Nowak [1981a] 2011, p. 250)

Nowak's views were expressed even more vividly in his answer to the question about the understanding of the political situation by the intellectual elites in the summer of 1981:

> Generally, the existing situation was not understood at all. In my opinion, the propositions for discussion within Solidarność constitute a terrifying discourse. They are like a dish of four scrambled eggs: patriotic, democratic, Christian and socialist. A social doctrine must be simple and be founded on the acknowledgement of the conflict and the place a given society occupies. These characteristics are nowhere to be found in the propositions. In short, the masses will rise to the occasion in an unusual manner, while the intellectual elites have failed in an unusual manner. There is not a single trace of a serious social policy fit for the existing situation. (Nowak [1981d] 2011, p. 273)

The work organization of the policy committees at the 1st NSZZ Solidarność National Convention was conducive to the development of eclectic policies. In Nowak's view:

> Policy was entirely in the hands of experts. I am one of them. We can imagine about 20 people sitting together, some of them representing completely opposite views. Agreement was reached only by cutting off

differences. What remained were clichés that everybody agreed to. ([1981h] 2011, p. 305)

One of the reasons for such a state of negotiations was the lack of distinction between the social roles of an expert and a thinker. The former's task is to work out technical details, while the latter's task is to develop ideas. Experts will never replace ideologues. Nevertheless, the mission to frame a new platform was given to experts, which then resulted in the lack of social ideas essential for any social movement.

The criticism of the political thought of the Solidarność opposition became more strident in the 1980s. Nowak agreed with Ash who observed that the Solidarność political platform was an implementation of the famous essay "How to be a conservative-liberal socialist?", in which Leszek Kołakowski ([1978] 1984, p. 205) – by way of enumerating features of conservative, liberal and socialist thinking – regarded the non-contradiction of these features as the sufficient condition of their merger. According to Nowak, the non-contradiction criterion was a necessary but insufficient condition for the merger of assertions from various ideological conceptions. The factual criterion of choice of some specific assertions from different ideological conceptions was somewhat omitted by Kołakowski and thus his selection was fairly arbitrary. Nowak noted sharply that Kołakowski's conception was "a counting-out rhyme rather than a construct" (Nowak [1985b] 2011, p. 601).

Leszek Nowak differentiated, however, between intellectual chaos or anarchism and ideological pluralism:

> In ideological pluralism, serious thinking alternatives are mutually incompatible, but they are always serious, i.e. they develop some specific points of view by exploiting their explanatory capacities as far as possible. Selective picking is for politicians, not theorists. ([1985b] 2011, p. 602)

Nowak recognized such clear conceptions as liberalism or Trotskyism, with which he polemicized in the years 1981-1989, but not eclectic patchworks of various ideological concepts.

2.2. *On Cognitive Deficiencies of the Solidarność Political Thought*

The cognitive deficiencies of the Solidarność political ideas resulted, in Leszek Nowak's view, from two processes: the ideological impact of Marxism on independent social awareness and inaccurate recognition of the nature of socialism by the opposition thinkers.

On the basis of his theory, Nowak argued that, in the West, the cognitive value of classical Marxism gradually decreased since this particular doctrine failed to grasp the processes of weakening the class struggle in the second half of the 19th century and the first half of the 20th century as well as the processes of accumulation of power and property in the second half of the 20th century. In socialist countries, Marxism fulfilled an ideological role and falsely explained the seizing of production property by the state authorities as socialization, and "class struggle" as struggle against enslavement. However, after 1956, the cognitive quality of Marxism slowly increased since workers' protests in Berlin (1953), Poznań and Budapest (1956) could be interpreted as class struggle (in the economic way). Nowak observed that:

> It was to the regime's best interest to remove Marxism with its views of antagonism, class struggle and solidarity of the oppressed from the spectrum of ideas. The banners of solidarity would then become more significant in Poland after August 1980: solidarity of the nation and solidarity of the state. Suddenly, the Marxist perspective concerned with the search for conflict of interests and viewing the oppressed as a source of social progress became replaced by the solidarity perspective in which the state was viewed as an all-encompassing organism in which social problems are only of moralist and praxeological nature. (Nowak [1981b] 2011, p. 129)

Moreover, as though through some "childish response," the ideological dominance of Marxism discredited all kinds of materialistic thinking in the opposition circles (Nowak 1981c [2011], p. 320). In Nowak's opinion, materialism was still a promising thought, provided that Marx's explorations were overpassed and developed, not ignored and rejected. In the political opposition thought in the years 1976-1980, but also later, in the policy program and political ideas of NSZZ Solidarność, a revival (in various degrees) of all types of political ideas could be observed: Christian, national, social democratic, liberal, conservative. Nowak did not negate the interests in the history of one's own and others' political ideas, provided that one had come up with some novel thinking earlier:

> Such references can be highly beneficial. They may reveal the paths of development of social thought, they may emphasize divisions, they may reveal unexpected similarities. Those historians who are aware of research methodology know well that the sources contain only what the researcher's imagination has already captured. Purposeless digging in the archives is merely a pretense of political thinking, which allows to conceal the researcher's lack of ideas and courage. (Nowak [1985a] 2011, p. 605)

Nowak warned against the consequences of ideological eclecticism:

> Eclectic political programs are only for patching up differences, gathering people who share nothing in common but plain negation of the system. Such a gathered mass cannot act in a uniform way; it can last and wait for

a knockout. It can also eliminate pluralism. If all these gathered people pretend to be unanimous because they have combined a few assertions, they have certainly eliminated their extreme differences and constitute an ideologically non-committed mass. These are social consequences of eclecticism in social thinking. This is what we experienced in Solidarność. ([1985b] 2011, p. 605)

He also pointed to the social reasons behind the lack of ideological novelty:

> Thinking more requires personal courage. In modern Poland people are more afraid to be alone among their own than to be repressed by the state. This can only explain the astonishing phenomenon of gregarious instinct in modern thinking of the Poles. We are patching up differences only to keep up appearances of unity as we are afraid to go up against the flow. (Nowak [1985a] 2011, p. 605)

The aforementioned ideological doctrines were formulated in response to different social conditions and were incapable of proper interpretation of the system of real socialism since they were not properly adjusted by their adherents and propagators. For Nowak, simple efforts to merge or separate ideological doctrines in Kołakowski's style were not enough. Instead he postulated the development of new political thinking. In the preface to his unpublished work "Dogmas of liberal thinking," Nowak made the following suggestion:

> It is because people in Poland today need new ideas [. . .] and nobody has actually invented a better way of developing new ideas than liberal thought. We need as many people thinking in different and independent ways as possible to think in a new way, i.e. polemicizing with the hitherto firmly established ways of thinking, with the risk of errors and failures, and with no applause from those thinking the old way. There is no other way in today's Poland but to try to think in a new way. Foundations must be sought for new political thinking. (Nowak 1986, p. 1)

The basis for the new and original ways of political thinking should be an accurate diagnosis of the conditions of Polish society in the late 20th century. According to Nowak's non-Marxian historical materialism, real socialism is a system in which one and the same social class controls the economy, politics and culture. In this perspective, such a system was the most oppressive one as it was based on a three-fold class monopoly. The main line of division in this system lied between the people's class and the triple-ruling class termed "state authorities" or "party apparatus." The main interest of this class was the maximization of political domination power leading to a sharper conflict of interests than the maximization of profit. Moreover, its control over the economy and culture even further contributed to its dominance over the rest of society. Nowak compared his criticism of real socialism based on non-Marxian historical materialism with the criticisms made from other ideological perspectives: liberal, revisionist and orthodox Marxist.

The most obvious and common perception in NSZZ Solidarność of the state system was the one based on such key categories as "credibility," "state" and "society." In Nowak's view, the dominant thought perspective in Solidarność consisted of the following three interlaced assertions:

> (1) Both components, state and society, are indispensable for sustaining human communities. The problem is that the state sometimes attempts to subjugate the society, in which case it becomes totalitarian. The society in its resistance against this subjugation sometimes abandons the state organization and transforms itself into an anarchic mob.
> (2) A nation, i.e. a community composed of the authorities and society, exists in a system of bilateral compromises. Social order is attained through citizens' control of authoritarian ambitions of the state and through self-restraining of civil masses for the common good.
> (3) The system of compromises can be durable under one condition: the rulers must be credible to the society. Only if society trusts the rulers can conflicts be solved by way of negotiation and serve a useful purpose, i.e. a signal indicating the need of adjustment to the system. Without credible authorities, the conflicts of interests will ruin the state, which is also indispensable to the society. (Nowak [1981f] 2011, p. 329)

These assertions, according to Nowak, refer to 19th-century liberalism, which emerged in response to totally different social conditions, i.e. free market capitalism, in which the state was playing the "night watchman" role. The emphasis in liberalism on the significance of institutional aspects of public life as well as advocating the introduction of free elections, political multiparty system and other forms of state institutional control was perhaps effective in 19th-century capitalism, where the social power of political authorities was counterbalanced by the existence of private property and independent public opinion. This emphasis, however, failed to function in the conditions of real socialism. The liberal ideology failed to notice that 'mono-party' or 'state' were simply cover terms for a triple-ruling class of rulers, owners and priests. The institutional channels of control of the triple-rulers are too weak and they should be preceded with a transformation of the triple-ruling class into a single class of rulers devoid of disposal of the means of production and the means of mass communication.

In Nowak's views ([1981e] 2011, p. 201; [1983b] 2011), the conception of human and civil rights derived from liberalism and present in policy enunciations of the NSZZ Solidarność was almost no different from the revisionist criticism of socialism. The latter was based on the category of alienation in the writings of young Marx, which meant that an individual's loss of control of man-made products would lead to the alienation of the individual in the social world. In the socialist system, the source of such alienation was state bureaucracy. For Nowak, both conceptions pointed to the same phenomenon:

> A good state respects an individual's aspiration, and if the state becomes an evil one the society restores its control over the state and the state becomes anew what it really is: an indispensible social tool for the community. (Nowak [1981e] 2011, p. 202)

According to Leszek Nowak, the weakness of the revisionist criticism of socialism lied in its renunciation of the class perspective and the specificity of socialism. In socialism, alienation can affect an ordinary party apparatchik, who fails to control the rivalry between individual party factions, as well as an ordinary citizen controlled by the triple-ruling class. Moreover, the notion of alienation is too insufficient a tool to describe social relations in socialism. The fundamental relation in socialism is enslavement in which citizens renounce their own preferences (their own will) and follow others'. Alienated individuals in a social system do not control the system, i.e. are subject to their own products, but still retains their own will. Enslaved people renounce their own will. Nowak described the relations between enslavement and alienation in the following way:

> An enslaved individual in a certain social system is alienated in this social system, but never the other way around. One can be alienated and still retain free will and control over one's actions. (Nowak [1983a] 2011, p. 212)

Nowak ([1981b] 2011, p. 126) had relatively the highest respect for the orthodox Marxian criticism developed by Milovan Dilas (1958), in which the party nomenklatura was referred to as a collective owner using state violence to maximize surplus product and to stimulate economic development. The state control of the economy and the role of the state in forced industrialization resulted from the necessity to modernize Eastern European countries and to catch up with civilizational achievements of better developed Western countries. Nowak criticized this perspective for being a case of historiosophical economism sustaining the myth of "the developed West," and for perceiving class divisions only in the sphere of economy. The Marxian criticism of real socialism failed to notice that the party apparatus was constituted by a distinct type of social interest (maximization of power regulation), which gave rise to wider social discrepancies that the maximization of surplus product.

On the basis of the aforementioned analysis, it can be suggested that perhaps one of the intellectual origins of the ideological eclecticism of the Solidarność political thought were the cognitive shortcomings of liberalism, Catholic personalism, or revisionism evolving towards social democracy. The insufficient recognition of the nature of real socialism by a given doctrine led to the piecing out of these shortcomings with concepts and categories adopted from other ideological and theoretical orientations.

2.3. *On Policy Deficiencies of the Solidarność Political Thought*

The strikes in July and August of 1980 and the foundation of Solidarność were perceived by Nowak as a yet another people's revolution against the system of triple authority. However, the effectiveness of this revolution depended on its accurate recognition of the opponent and on its adequate methods of fight. The inaccuracy of recognition of the nature of real socialism noted by Nowak led to defective policy proposals. The social thought of Solidarność evolved from the role of "trade union supervisor of the authorities" overseeing the implementation of the social accords, to the role of "co-decision-maker" (for the periodization of policy evolution of the NSZZ Solidarność, see Brzechczyn 2010, p. 25). Nowak opposed both roles. At the beginning of the Solidarność revolution, he assessed the prospects of Solidarność in its clash with the triple-rule system:

> The real prospect is that the unions will become social unions rather than trade unions. The real danger is that they will not become social unions and their social role will not be assumed by other institutions such as councils or self-governments. A particular threat is that the unions might turn out to be simply a single, centralized hierarchical organization. If this happens, if a socialist structure is recreated within the unions, i.e. a division into decision-makers and sycophants is reintroduced, then it will be quite irrelevant whether this structure is independent of the communist party and whether union activists have good intentions. The new elite will be sooner or later absorbed by the bureaucratic milieu. (Nowak 1980a, pp. 9-10)

Nowak also opposed the transformation of Solidarność into a political party (parties) and its entry into the structure of power. He argued that:

> If Solidarność joined, directly or by supporting some groups or institutions, the structure of political power, it would become a component in the triple rule system. The economic power of the communist party apparatus has not been dissolved yet. Party committees in industrial plants are awaiting some better circumstances in which they could become active again. The economic sections of the higher levels of the apparatus do not even make it secret that they want to revive the economy in their own ways by concentrating decisions in their own hands like back in the old days. The

doctrinal power of the party apparatus is still there. One can see it clearly in the press or on television with each mentioning of the cold war against Solidarność. If Solidarność entered the structure of power even indirectly, it would mean the acceleration of bureaucratic degeneration (already quite visible, as I have written many times before) and a fusion with the system of triple-rule system. One must not even think about participating in power until the owning state and the doctrinaire state becomes a normal state. (Nowak [1981f] 2011, p. 334)

In the fundamentalist portfolio co-authored by Leszek Nowak and presented to the 11th Policy Team "Solidarność towards the Polish United Workers' Party and the State Authorities," NSZZ Solidarność was to return the means of production and the means of mass communication to the people (Nowak [1981g] 2011). Nowak also advocated a radical self-government reform, in which workers' self-managements would be given the right to appoint directors, and he criticized the compromising stance of Solidarność on self-government, which, in his opinion, led to a number of possible local conflicts (Nowak [1981f] 2011, pp. 332-333).

A synopsis of Nowak's criticism of the Solidarność political thought was an essay "A Few Theses on Contemporary Polish Society" ([1983b] 2011) written during the martial law period in Poland. Nowak described the Solidarność movement in the years 1980-81 as the greatest revolution in the triple-rule system with a false awareness unfit for the existing circumstances. The key manifestations of this false awareness included:

- solidarism – a conviction that national ties are stronger than national divisions (manifested, for example, by negotiations following a partner format ("like a Pole with a Pole"), and not the model of people's class representative against the triple-lord);
- idealism – the source of power lies in social credibility; however, the communist party apparatus lost the source of its ideological legitimacy but did not relinquish its control over the means of coercion;
- institutionalism – instead of creating accomplished facts and initiating the transition of economic power by self-governments, Solidarność kept on negotiating with the communist state authorities and sought compromise.

Conclusion

Despite the fact that the mass Solidarity movement in Poland in the years of 1980-81 was a decisive impulse which led to the ultimate collapse of real socialism in Poland and Eastern Europe a decade later, its ideological legacy and, more importantly, social praxis in the years 1980-81 affected the course and the shape of social transformations in Poland only slightly. This paradox has been noted by a number of Polish and foreign authors representing often different ideological orientations. According to Arista Maria Cirtautas:

> At first glance, it would appear that institutional development in Poland since 1989 has little to do with either the ethos or the experience of Solidarność. The goals advocated by the movement such as self-government and workers' self-management seem outdated and irrelevant in the context of competitive party politics and privatization. (Cirtautas 1997, p. 206)

David Ost even wrote about a bitter failure of the Solidarność ideals:

> Isn't it a blight on postcommunist democratization that the chief losers were those who made the transformation possible? That those whose solidarity strikes had helped make capitalist democracy possible would soon find themselves working in firms whose managers tolerated neither unions nor collective bargaining. (Ost 2005, p. 17)

Zdzisław Krasnodębski, on the other hand, perceives Solidarność as an unfinished project that was aimed at restoring freedom, equality and solidarity. Freedom in this project was understood as the "self-management of citizens through their participation in public life and through collective self-determination" (Krasnodębski 2006, p. 114). Equality was understood as equality before the law, equal dignity of all as well as equality of opportunity and vote. In the economic sphere, Solidarność intended to improve Polish labor and make Polish citizens owners. The social project of Solidarność was "a project of democracy built by the nation and for the nation – the nation understood as open and friendly to others, solidary community of tradition, culture and history bound together by values, memory and symbols" (Krasnodębski 2006, p. 117). In Krasnodębski's view, in Poland after 1989, freedom has been perceived as a possibility to realize one's own preferences regardless of the needs of the community. Poland is still a country witnessing civilizational divisions and inequalities between particular regions of the country and layers of society. The owner has been replaced by the swindler.

In fact, multiple factors responsible for this situation can be found (see, for example, Brzechczyn, 2004; 2011a; 2011b, pp. 173-178). It seems that one of the intellectual factors that affected the renunciation of the ideological legacy of Solidarność was the intellectual provincialism of the movement, reflected in its inability to challenge the established stereotypes of political thinking. The ensuing ideological eclecticism, i.e.

invoking a few ideological traditions at the same time, brought about confusion as to which of the ideologies should be modified and developed in changing social conditions. As a result, none of the ideological traditions was developed and at the turn of the 1980s and 1990s Solidarność simply pandered to the least controversial and the most common Western European ideology, i.e. neoliberalism, which began to dominate the political discourse in Poland.

Provincialism in Polish political thinking made even the few elements of novelty in the Solidarność message fail to find an ideological continuation in the political transformations after 1989. They were simply rejected and ignored but not critically surpassed. At the threshold of the transformation, when Leszek Nowak was asked about potential threats in the process of democratization, he said:

> In the intellectual dimension, an old Polish hazard hides. The Poles are physically a very courageous people. There were no better lancers than the Poles. Unfortunately, we are unbelievably cowardly intellectually. There are no worse intellectuals than the Poles. Perhaps, I am exaggerating, but the tendency is visible. This has always been Witold Gombrowicz's problem: What to do if the Poles are on the periphery of world culture? Should they move toward the center by all means, becoming only a transmission belt of culture from the center to the periphery? Or should the Poles overcome their indigence by, first of all, admitting it (which can be difficult) and, second of all, by trying to find answers to the situation they have found themselves in. I suppose the consequences of intellectual cowardice can be very serious to the Poles, since an intellectual coward is only bold enough to follow those who are culturally stronger, i.e. the West, and to replicate concepts devised at some different time and in conditions that could have been different from the Polish ones. (Nowak [1989] 2011, p. 705)

In my opinion, the above diagnosis is still valid today in the age of new phenomena (the Internet, globalization) and challenges (socio-economic crisis in the United States and the European Union), and it should make the Polish social and political elites reconsider the position of Poland in the contemporary world and the ways to pursue Polish national interests.

Adam Mickiewicz University
Institute of Philosophy
ul. Szamarzewskiego 89C
60-568 Poznań
Poland
e-mail: brzech@amu.edu.pl

REFERENCES

Arendt, H. (1963). *On Revolution.* New York: The Viking Press.

Ash, T.G. (1983). *The Polish Revolution. Solidarity*, London: Jonathan Cape.

Brzechczyn, K. (2004). The Collapse of Real Socialism in Eastern Europe versus the Overthrow of the Spanish Colonial Empire in Latin America: An Attempt at Comparative Analysis. *Journal of Interdisciplinary Studies in History and Archaeology* 1 (2), 105-133.

Brzechczyn, K. (2010). Program i myśl polityczna NSZZ "Solidarność" [The Program and Political Thought of NSZZ "Solidarność"]. In: Ł. Kamiński and G. Waligóra (eds.), *NSZZ "Solidarność," 1980-1989*, vol. 2: *Ruch społeczny*, pp. 13-74. Warszawa: Instytut Pamięci Narodowej.

Brzechczyn, K. (2011a). The Forgotten Legacy of Solidarność and Lost Opportunity to Build a Democratic Capitalist System Following the Fall of Communism in Poland. In: N. Hayoz, L. Jesień and D. Koleva (eds.), *Twenty Years after the Collapse of Communism. Expectations, Achievements and Disillusions of 1989 (Interdisciplinary Studies on Central and Eastern Europe*, vol. 9), pp. 395-416. Bern: Peter Lang.

Brzechczyn, K. (2011b). Freedom, Solidarity, Independence: Political Thought of the "Fighting Solidarity" Organization. In: D. Dobrzański (ed.), *The Idea of Solidarity: Philosophical and Social Contexts (Cultural Heritage and Contemporary Change. Series IVA, Eastern and Central Europe*, vol. 42: *Polish Philosophical Studies*, vol. 10), pp. 159-179. Washington DC: The Council for Research in Values and Philosophy.

Cirtautas, A.M. (1997). *The Polish Solidarity Movement. Revolution, Democracy and Natural Rights.* London & New York: Routledge.

Ciżewska, E. (2010). *Filozofia publiczna Solidarności [The Public Philosophy of Solidarność].* Warszawa: Narodowe Centrum Kultury.

Dudek, A. (2002). Rewolucja robotnicza i ruch narodowowyzwoleńczy [Worker's Revolution and National and Liberation Movement]. In: D. Gawin (ed.), *Lekcja Sierpnia. Dziedzictwo "Solidarności" po dwudziestu latach*, pp. 143-158. Warszawa: IFiS PAN.

Dżilas, M. (1958). *Nowa klasa [The New Class].* Translated by A. Lisowski. Warszawa: Oficyna Liberałów.

Kołakowski, L. ([1978] 1984). Jak być konserwatywno-liberalnym socjalistą? Katechizm [How To be a Conservative-Liberal Socialist? A Catechism]. In: *Czy Diabeł może być zbawiony i 27 innych kazań*, pp. 203-205. London: Aneks.

Krasnodębski, Z. (2003). *Demokracja peryferii [Democracy of Peripheries].* Gdańsk: Słowo/obraz terytoria.

Krasnodębski, Z. (2006). Niedokończony projekt [An Unfinished Project]. In: *Drzemka rozsądnych*, pp. 113-119. Kraków: Ośrodek Myśli Politycznej.

Nowak, L. (1973a). Cnoty uczonego [The Virtues of a Scholar]. *Nurt* no. 5, 24.

Nowak, L. (1973b). Dowcip jako kryterium prawdy [Joke as a Criterion of Truth]. *Nurt* no. 12, 15.

Nowak, L. (1973c). Istota odkrycia naukowego [On the Essence of Scientific Discovery]. *Nurt*, 1973, no. 7, 20-21.

Nowak, L. (1973d). Nauka i zdrowy rozsądek [Science and the Common Sense]. *Nurt* no. 4, 1-2.

Nowak, L. (1973e). Teoria inteligentnego dyletanta [A Theory of an Intelligent Dilettante]. *Nurt* no. 6, 30.

Nowak, L. (1973f). Teoria wobec nowych faktów [Theory and New Facts]. *Nurt* no. 10, 12-13.

Nowak, L. (1974a). Doktryna i prawda [Doctrine and Truth]. *Odra* no. 6, 19-22.

Nowak, L. (1974b). Eclecticism and Tolerance. *Revolutionary World: An International Journal of Social Philosophy* 8, 91-99.

Nowak, L. (1974c). Nauka i młodość [Science and Youth]. *Nurt* no. 11, 14-15.

Nowak, L. (1980a). Głos w dyskusji [A Voice in the Discussion]. In: *Lato 1980. Dyskusja.* Poznań: WiW.

Nowak, L. (1980b). *The Structure of Idealization. Towards a Systematic Interpretation of the Marxian Idea of Science. Synthese Library*, vol. 139. Dordrecht/Boston/London: Reidel.

Nowak, L. ([1981a] 2011). Cena braku perspektywy [The Price of the Lack of Perspectives]. In: Nowak (2011), pp. 250-253.

Nowak, L. ([1981b] 2011). Marksizm dzisiaj. Przyczynek do kategorialnej historii marksizmu w systemie trójpanowania klasowego [The Marxism Today. A Contribution to the Categorial History of Marxism in a Triple-Class System]. In: Nowak (2011), pp. 117-130.

Nowak, L. ([1981c] 2011). Nasz ruch musi być inny . . . [Our Movement Has to Be Different . . .]. In: Nowak (2011), pp. 318-321.

Nowak, L. ([1981d] 2011). O fundamentalnym błędzie Marksa, totalitaryzmie, istocie stalinizmu, "trójpanach" w PZPR, tezach "Solidarności" i szansach społeczeństwa bezklasowego [On the Fundamental Error of Marx, on Totalitarism, on the Essence of Stalinism, on 'Triple-Lords' in PZPR, on the Theses of Solidarność and the Chances of a Classless Society] In: Nowak (2011), pp. 271-277.

Nowak, L. ([1981e] 2011). Socjalizm rodzi nie alienację, lecz zniewolenie [Socialism does Not Bring about Alienation but Enslavement]. In: Nowak (2011), pp. 201-204.

Nowak, L. ([1981f] 2011). *Umiarkowanie w myśleniu prowadzi do nikąd* [Moderation In Thinking Leads Nowhere]. In: Nowak (2011), pp. 329-335.

Nowak, L. ([1981g] 2011). Związek a PZPR i władze państwowe. Propozycje do programu Solidarności ["Solidarność" towards the Polish United Workers' Party and the State Authorities. Program Proposals]. In: Nowak (2011), pp. 316-317.

Nowak, L. ([1981h] 2011). Związek potrzebuje społecznych idei [Trade Union Needs Social Ideas]. In: Nowak (2011), pp. 305-306.

Nowak, L. ([1983a] 2011). Alienacja i zniewolenie. Szkic o zasadach krytyki socjalistycznej formacji społecznej [Alienation and Enslavement. An Outline of the Rules of Criticism of the Socialist Social Formation]. In: Nowak (2011), pp. 205-213.

Nowak, L. ([1983b] 2011). *Kilka tez o współczesnym społeczeństwie polskim* [A Few Theses on Contemporary Polish Society]. In: Nowak (2011), pp. 397-411.

Nowak L. (1983c) *Property and Power. Towards a non-Marxian Historical Materialism Theory and Decision Library*, vol. 27. Dordrecht/Boston/Lancaster: Reidel.

Nowak ([1985a] 2011). Bogactwo i nędza trockizmu. Kilka słów komentarza [The Richness and the Poverty of Trotskyism. Some Comments]. In: Nowak (2011), pp. 605-608.

Nowak ([1985b] 2011). Od reformistycznej do rewolucyjnej teorii socjalizmu [From Reformist to Revolutionary Theory of Socialism]. In: Nowak (2011), pp. 590-604.

Nowak, L. (1986). Przedmowa [Foreword]. In: *Dogmaty myśli liberalnej*, pp. 1-2. Poznań: Manuscript.

Nowak, L. ([1989] 2011). O porozumieniu Okrągłego Stołu, uwłaszczeniu nomenklatury, perspektywach anarchizmu i polskim prowincjonalizmie [On the Round Table Agreement, the Enfranchisement of Nomenklatura, Perspectives of Anarchism and Polish Provincialism]. In: Nowak (2011), pp. 700-706.

Nowak L. (1991) *Power and Civil Society. Towards a Dynamic Theory of Real Socialism.* New York/London: Greenwood Press.

Nowak, L. (2011). *Polska droga od socjalizmu. Pisma polityczne 1980-1989* [*Polish Road from Socialism. Political Writings 1980-1989*]. Poznań: Instytut Pamięci Narodowej.

Nowak, L and I. Nowakowa (2000). *Idealization X: The Richness of Idealization. Poznań Studies in the Philosophy of the Sciences and the Humanities*, vol. 69. Amsterdam & Atlanta: Rodopi.

Ost, D. (2005). *The Defeat of Solidarity. Anger and Politics in Postcommunist Europe.* Ithaca, NY: Cornell University Press.

PART IV

GLOBALIZATION, SOCIETY, CULTURE

Max Urchs
Uwe Scheffler

PARADIGMS, MARKETS, AND POLITICS

FROM PROVINCE TO METROPOLIS AND RETOUR[1]

ABSTRACT. In times of modern information technology, the world of science is becoming smaller. Does this mean that there will be no more provinces? We do not think so. Setting out from Leszek Nowak's thought "province is where one thinks not on one's own account but on account of another," we indicate a number of processes (both internal and external to the sciences) that perpetuate provinces. These processes are driven by specific access to scientific knowledge, by education, by new forms of communication, by shortage of financial support and the concentration of resources. We look at the interplay between criteria of theory choice and location on the scientific map. Next, we explore the connection between geo-social and scientific provinces, taking into consideration political and cultural parameters. The conceptual framework of metropolises and provinces in science turns out to be, though not all-embracing, an extremely fruitful one.

1. Provinces and Metropolises: Exploring the Dialectic

The world is becoming smaller. One may certainly feel indifferent about this fact, but it is a fact. For us, San Francisco is as far from Berlin as Poznan was for our ancestors 200 years ago – some 12 hours to travel. We knew everything about the Fukushima disaster almost immediately, except maybe that which the authorities did not let us know. Talking to a friend on the opposite side of the globe is not only possible – it is easy, cheap, and reliable. Of course, there are places on Earth far from an airport or train station, unconnected to the World Wide Web, without a decent television broadcast or mobile network. Often they are perceived as a

[1] We are very much indebted to the editors of the volume for their enduring support and most helpful comments.

In: Krzysztof Brzechczyn and Katarzyna Paprzycka (eds.), *Thinking about Provincialism in Thinking (Poznań Studies in the Philosophy of the Sciences and the Humanities*, vol. 100), pp. 237-258. Amsterdam/New York, NY: Rodopi, 2012.

rearward province in the common sense of the word: uncultured, unsophisticated, and backward. But there are fewer of these waylaid places now than there were many years ago, and hopefully, one can assume that there will be even fewer in the years to come. Technically, every single person will probably be able to partake in the wealth available in our world even now.

In the sciences and the humanities, this smallness of the world has become even more evident. From a university office, many high profile international journals, sometimes even books, can be accessed without leaving one's chair. Writing or discussing a paper with colleagues from different universities all over the world is a common activity. Conferences are held that include talks given via the internet thousands of miles away. Interestingly, in contrast to the geographically small world, the world of the sciences and the humanities seems, at first glance, not to have provinces at all. Provided that scientific research is carried out at academic institutions, scholars are trained at universities, and those scholars are usually affiliated to universities, there will be no single centers in academia without electronic journals, internet or e-mail contact. Agreed, the access may differ in quality and speed but this is improving rapidly – much faster than among daily living conditions. Does that mean there are no provinces in science and humanities? We do not think so. Although there is certainly a tendency toward de-centralization and internationalization of the academic world, there still are metropolises and provinces in science and humanities. Moreover, we will show that there are a number of processes, both internal and external to the sciences, which are reproducing provinces over and over again. The emerging picture thus seems inconsistent. On the one hand, we see the World Wide Web as the great leveler, bulldozing down all differences between scientific institutions. On the other hand, there are economic and political processes as well as processes internal to the sciences, which dialectically change yet perpetuate the structure formed by province and metropolis with respect to science. Inconsistencies attract attention. So let us give the topic a closer look.

Thinking about provincial behavior in the sciences, the first association is perhaps mental narrowness, indolent or routine thinking, a lack of courage to reach for fresh thoughts. In a word: to think provincially means to intellectually stew in one's own juices. In English, such a state of mind is usually called insularity. That concept, however, does not cover at least one important overtone of provincial thinking. Provincial scientific thinking – meaning insular, isolated scientific thought – is an oxymoron. That seems to be true for English but not for all languages. In German, as well as e.g. in Polish, "provincial thought" is a consistent and comprehensible concept also with regard to science. What is missing in "insularity" is an asymmetry of intellectual influence: metropolitan thoughts have an impact on provincial ones, but not vice versa. Consequently the dictionary definition of a province as "a territorial possession controlled by a ruling metropolis" does not fit in scientific context. Metropolises are barely interested in what goes on in the provinces. In particular, not much control is extended.

Leszek Nowak in one of his papers put forward a novel facet to this: "province is where one thinks not on one's own account but on account of another" (Nowak [1998] 2012, p. 65). A fascinating idea, indeed! Setting out from that thought, Nowak proposes a clear criterion that should locate province on the mental map of the sciences: just check who is citing whom. We will thereby find, so his thesis goes, that the direction of citations (its asymmetry) yields a hierarchical status of the specific scientific groupings. In terms of such a hierarchy, the author who is cited dominates the one who cites. The one who stands at the very top of the citation-driven hierarchy determines where the metropolis is located. It is there, of course, where he himself stands, or rather sits. Whoever lives at the end of the citation chain is provincial. Unless he is lucky enough to belong to a scientific community whose leading exponents, by their creative power, raise the whole school into a scientific metropolis. We will return to Nowak's proposal in §2.

Historically, the first step in liberating the access to knowledge was probably the invention of writing and archiving of the written text on papyruses some 5000 years ago. Before that period, obtaining knowledge was bound to the place where a teacher was. After that, all kinds of written knowledge could be transported. The very fact of exclusively oral teaching made sure that astronomical, mathematical or philosophical knowledge was essentially centralized. Even in well developed countries almost everyone lived in intellectual provinces. Educated people very often had to travel a great deal more to establish contact with other people of their rank. Books' lifespans can extend over years and years, being copied over and over again. What is more, they can be sent to other places. People cannot. Though a scribe certainly belonged to the educated, it was much easier to copy books than to write them. It was books that in post-oral times institute contact and provide a link between scholars.

The concept of a province makes sense only when comparing it to that of a metropolis. The process of creating scientific metropolises was already set out during the time of oral teaching through the building of schools. Just because the intellectual enterprise was local, some scholars and their students had better knowledge, while others did not. Albeit, travelling story tellers, singers, preachers, and teachers might have played an important role – their numbers were certainly not high enough to ensure area-wide education. Moreover, teachers aligned to specific traditions, which were imparted in small groups.

Written texts, however, once in use, created new intellectual metropolises, leaving most of the provinces behind. Books were read where they were copied, and *vice versa*. They were rare and expensive. The emphasis is, of course, on the big libraries. How big a library could grow demonstrated how intellectually important the place was, and, to a large extent, it was a question of available funding. The connection between economic success and intellectual development, already present in the oral tradition, became even more intertwined in times of creating books. The people involved in reading and copying, all the monks, scholars, and at least some of the students had to be fed, clothed, accommodated, and perhaps paid by someone. Naturally, there were only very few intellectual centers.

There were two other important steps as well. First, there was formalization and regulation of university education about 1000 years ago. Second, more than 500 years ago, the appearance of printed books provided relatively cheap, multitudinous, and thus by far more easily available texts. The former established new metropolises of knowledge – Paris, Cambridge, and Cracow among them. The latter made education more effective and available to a much larger group of people. Hence, while book printing did not put an end to provinces, it made living in a province far more bearable. In about the same times, i.e. the Renaissance, scholarly trained adventurers with good hand writing travelled to the back provinces north of the Alps, copying or stealing ancient Greek texts that had survived in devastated monastic libraries. Written texts thus transported knowledge to other places and to other times. Even if a flourishing intellectual metropolis had expired long before, its traditions might illuminate a new intellectual metropolis somewhere else (see Greenblatt 2011).

During the last 200 years, capitalist economics, technological needs, and, finally, democratization have influenced the relationship between intellectual provinces and metropolises. We will scrutinize that topic in §3 and §4. Today, worldwide cooperation between scientists is always possible. There is no need to work at the same university or live in the same city or country. The World Wide Web is flattening the intellectual landscape considerably. Is there a possibility for the province-metropolis distinction to dissolve altogether? Most likely that will not happen. We elaborate on our opinion in §5.

In the context of science, the concept "province" is implicitly used in two non-identical meanings. The first meaning takes province to be small and under-funded, and the second one just indicates that it is not a part of the paradigm-setting group. Clearly, the first one refers to institutions, e.g. universities or institutions, whereas the second one applies to organizations, i.e. researchers or research groups. There are good reasons to believe that the modern system of conducting science and humanities reproduces provinces in both senses.

2. Intrinsic Drivers

We identified several processes that determine the dynamics of the metropolis-province relationship. First, access to knowledge, education, and communication blur the boundaries. This is something that modifies and shapes all the other processes, and it is nowadays driven by internet technology. Second, social consensuses, practical needs and the desire to develop or sometimes compete with neighbors, rivals and friends bring about institutions at prestigious locations. Third, shortage of financial support and their concentration creates well-funded towers in an otherwise flat landscape. Fourth, social and intellectual membership in groups with similar educational and scientific background creates importance. The latter process is the subject matter of theory choice.

What is the criterion for scientific importance? Nowak ([1998] 2012) discerns three types of scientific personalities among researchers: the creative mind, the critical one, and the applicational personality type. The first one has brilliant new ideas that propel the entire scientific discipline forward and are attractive enough to be considered by other scientists. The critical mind is not able to come up with ideas comparably rich in content, yet their value is understood and they are intellectually independent enough to critically examine them. In the process, they manage to improve them, correct errors, etc. Researchers of the third type are limited in their scientific capability to make more or less schematic use of the new ideas in solving open problems. They just carry on with normal science, in Kuhn's sense. In any scientific school, all three types of scientists are represented. Communities of scientists are provincial if they do not include the first type of research personalities, the creative minds. This thought of Nowak's, which emanates originality and straightforwardness, provides a good portion of solace and relief for all those who believe themselves to dwell in intellectual province. Scientific provincialism is, according to Nowak, not a geographical concept. A province in science is not a secluded region, to which the individual scientist is indefinitely banished. The scientist has the power within himself to free himself from the mental province. It is up to him to reach the metropolises of science.

It may seem that choosing a scientific community to belong to is nothing more than choosing a favorite theory, something which is dealt with by Kuhn (1969). Why should one support this theory instead of that one? In an influential paper (Kuhn 1977), Kuhn describes theory choice (for instance, whether to belong to the Einsteinian or to the Bohr paradigm of Quantum Mechanics) as value-based. Kuhn identifies five basic values: accuracy, consistency, scope, simplicity, and fruitfulness. There is a bundle of problems connected with this approach, among them, why these values and not others? How do we interpret their behavior, if they pull or push in different directions? Who determines what should be understood by these words? According to his famous "no algorithm" thesis, even if two scholars agree on the standards for all of these values, they must not necessarily opt for the same theory. Otherwise, as Kuhn stresses, there would be a solution to the problem of induction. He gives a telling metaphor: different theories become comparable like different languages, namely by translation. But then, we know, there might be many principally different translations, different manuals all of which are equally correct. In the same sense, there are many equally good ways of selecting a theory. Just as people do not choose their language based on rational criteria, their choice for or against a theory, and therefore a scientific community, is largely based on something else. It was Ludwik Fleck ([1935] 1993) who described science as a process that involves not only a mind and nature, but rather a mind, a team of minds (*Denkkollektiv*) and reality. Cognition, according to Fleck, has historical and interpersonal dimensions, which determine the content of scientific theories as well as their appeal and success. Why some scholars are part of a thought style (*Denkstil*), a paradigm in Kuhn's terms, and others are not is a result of a partially uncertain, highly chaotic process. Why some styles of thinking dominate and survive while others do not is a different question, but the answer to it, as well as to the former question, depends on many causes. So, not only the genetic questions of constructing a theory or subscribing to one, but also the systematic questions of justifying a theory, a community or one's belonging to it are not completely explicable in a precise way.

Both Fleck and Kuhn agree that it is impossible to predict when and where an important theory change will occur. Both agree on the idea that the roots of the new thought style, or paradigm, exist somehow – probably in the provinces of normal science. Both would agree on Nowak's idea that it requires a creative scientist to challenge the existing paradigm. This obviously explains why science and humanities develop in metropolis-province schemes: new theories start as singularities, as a result of research by a few people. These scholars have the advantage of setting the field, of publishing the first important papers. They themselves are a center of condensation, sometimes for a very long time. Quite often, they invent the appropriate vocabulary and train their own followers. Once taken over by normal science, a trend in the opposite direction occurs: more and more people work inside the new paradigm, the number of publications increases and students are trained in the field. Via the World Wide Web, collaboration is now fast and easy in all distances and the role of initiators is decreasing.

The sequence described above belongs to the level of scholars or groups of them. According to this view, science develops discontinuously, thereby creating metropolises, but not in the sense of institutions or even in the geo-social sense. It is the people who are provincial or metropolitan in this sense. Only in a derived sense can institutions, which house them, become more or less metropolitan by theory change and scientific progress. That does not mean that a paradigm change cannot be prevented by institutional or social conditions.

Such inner-theoretical activities fit clearly into Nowak's scheme of a province-metropolis relationship based on citation relations. A founder of a new paradigm should certainly be cited more often than others, simply because his was the groundbreaking work. Critical scientists, also often cited, answer the main questions and stake the claim. On closer inspection, Nowak's proposal does not seem unproblematic though. First of all, his analysis clearly focuses on the scientist's role as a researcher, while the figure of academic teacher merges into the background. In top-level papers, textbooks are seldom cited. It seems hard to assume that today's universities are populated by ideal scientists in Humboldt's sense, who contribute in equal measure to research and teaching. Therefore one would be better served not to abstract from taking academic teaching into consideration as well.[2] Second, it seems debatable whether the gradient of citations really displays the kind of asymmetry needed for generating a hierarchy. The overwhelming impression is rather an extensive ritualization of (mutual) citation, brought forward by the established methods of ranking in many scientific disciplines. In downright citation cartels, not only are the papers that need to be cited established, but even their order of appearance is determined. Any deviations are considered unkind and are threatened with restrictions. Consequently, there is no automatism that guarantees the presence of creative and critical thinkers at the top of the citation lists. Admittedly, the conventions described are

---

[2] Admittedly, teaching seems nothing to be proud of, nowadays. To teach less is a sign for scientific excellence. The most excellent scientists don't bother with teaching students at all.

relevant only between leading research groups. Only those who publish at
eye level make it to the relevant journals, the ones in which they
themselves are published and are relevant for the rankings. One may
assume anyway that these phenomena are not relevant insofar as they
pertain to research groups where the metropolitan status seems to be
beyond question. Nowak's proposal may well have its eligibility anyway.

The concept of scientific province, which arises out of bundles of
citation vectors, nicely supports another of Nowak's theses. In many areas
of sciences, there is a habit of massively citing currently fashionable
papers. In the above framework, such publications become big nodes of
the citation graph and form a new metropolis. In that sense, Leszek Nowak
is right in claiming that province in science has nothing to do with
location. Yet the big nodes are immediately identified by specialized
services. In times of low loyalty of scientists to their home university, top
researchers are highly mobile, especially when they hold an endowed chair
and the sponsor does not care about that specific university. Therefore,
wealthy institutions usually succeed in hiring the big nodes, i.e. the
according authors, shortly afterward. They regularly bring at least parts of
their research teams, or they immediately establish new ones. Successful
organizations thus tend to concentrate at wealthy universities. As a result,
the person-based citation graph aligns with the institutional metropolis-
province structure in science.

A truly prototypical example of the inextricable connections between
political factors and factors intrinsic to science is displayed by the fate of
the great Polish logician Stanisław Jaśkowski. Jaśkowski started his carrier
under supervision of the world-class logician Jan Łukasiewicz in Warsaw.
Warsaw was certainly not a dark scientific province at that time. The aerie
of the Lvov-Warsaw School was one of the centers of logicism, frequented
by recognized scientists from all over the world. Admittedly, logical
positivism was not the dominating current in European philosophy in those
days. With respect to language, work in Poland did not provide a
locational advantage. Consequently, Jaśkowski lost precedence for his
calculus of natural deduction to a German scientist, Gerhard Gentzen. That
had a lot to do with faster access to important, i.e. internationally visible,
journals. One of Jaśkowski's ideas concerned so-called paraconsistent
logic. Today, some of the most cited logicians, Graham Priest for example,
believe it to be a new paradigm in logic. Jaśkowski developed and
investigated one of the first calculi of this kind but the recognition of his
work was hindered by political and language reasons. By serendipity, in
the seventies of the last century, a Brazilian scientist – contrary to zeitgeist
– travelled to Eastern Europe, the province behind the Iron Curtain, to
search for new ideas in logic. Newton da Costa personally visited all the

places that may have been relevant for his own area of research, paraconsistent logic. Only that way did some of Jaśkowski's brilliant ideas enter into the worldwide scientific community. In that way, it was mainly da Costa, a logician from a Brazilian province, who secured Jaśkowski's international visibility. Naturally, he was not selfless enough to over-promote the project that competed with his own.

Chances that the reception of unknown results from the provinces will improve in the future are rather poor. First, the type of travelling scientist searching for ideas has become decidedly rare. And second, reading habits of scientists have changed. Papers that lie behind the temporal horizon designated by electronic publication just vanish. They do not appear in the normal search routines used by contemporary researchers. Regardless of their scientific value, they pass away to the Nirvana of the sciences. Well understood, the World Wide Web is Janus-faced: it provides easy access to modern results and fast publication, on the one hand, but, on the other, looking up older publications becomes more and more unusual and awkward. Certainly, forgotten treasures must still be waiting to be found on the outskirts of science. Yet barely anybody, except for a few historians of science, cares. Scientific production accelerates in tempo and size. Nowadays there seems to be hardly any demand in mining old ideas. The plethora of scientists all over the world brings out more new ideas than can be actually consumed. Paraconsistent logic, by the way, is recognized as part of a larger project leading from ancient dialectics to complexity theory. Even if one is, contrary to Priest, not ready to understand it as a new paradigm of logic, it is part of a developing paradigm of philosophy of science. Be that as it may, whenever scientific ideas become incorporated into common knowledge, they must not be ignored, regardless of their origin. The authors of an article "Sorting through the wealth of nations," working at the dignified Hoover Institution, reveal complete ignorance of inconsistency tolerant formal reasoning half a century after Jaśkowski, da Costa and others. This cannot be subsumed under the usual disinterest of the provinces. As an upshot, Hoover Tower is beamed into back province (without affecting the rest of Stanford's institutions).

Obviously, it is honorable and pleasant to be the founder of a new paradigm. Here, the personal metropolis status is connected with importance and promotion to metropolis-status institutions. Sometimes, the creation of a new theory is done with the help of essentially restricting the scientific field. As a result, the disciplinary structure of the academic organization dissolves by downright bizarre particularization of disciplines,[3] on the one hand, and by somewhat haphazard efforts to install transdisciplinary projects, on the other. An increasing compartmentalization is being mistaken for professionalism. Even philosophy as the all-embracing synopsis in the name of such a professionalism is driven into the analysis of more and more idiosyncratic details. The more divided into small section science is, the more paradigms are at hand, the more paradigm changes will occur and the more metropolises we will see. But they might be not that important anymore.

Paradigm change, the development of radically new theories, is therefore a permanent source of change in the existing province-metropolis relationship. It is permanent because, as far as we know today, science grows this way. There are times when province-metropolis differences shrink – this is normal science in Kuhnian terms. Contrary to all the tendencies mentioned at the beginning of §2, inner-scientific processes ensure that some hilly landscapes are created, but no-one can predict their specific topography.

3. Market Driven Processes

In times of only oral teaching, intellectual metropolises, if any, were connected with markets in the most literal sense.[4] Since poets, singers, philosophers had to make some sort of living, they competed for sponsors (first, hosts and later, paying customers) on the market place where many people got together. That was surely a double-edged strategy since better endowed places attracted more competitors. The problem remained fundamentally unchanged through the millennia. Today, creating a new theory or setting out a pioneering research project means not only having an idea and gathering the appropriate specialists. A large amount of time and work is devoted to finding appropriate sponsoring. Usually, economists approve of these mechanisms and point to competition as the main source and driver of progress – economic as well as scientific. As long as financial means are insufficient, there is a self-perpetuating, self-energizing circle: who disposes of manpower and funding will be more

---

[3] Jürgen Mittelstrass (1998) mentions hymnology as well as Brasilian language and literature studies.

[4] Beforehand, however, science had to create a market for the products of science. It was important to demonstrate that the new product, i.e. scientific knowledge, is useful or appealing, or both. And one had to establish an understanding that the author deserves payment for his intellectual work, though he does not lose his product by selling it in the same way as a tradesman does. In a wealthy polis that was much easier to achieve.

successful, and whoever is more successful will dispose of material means. The fact that the main centers of science and humanities, which used to be in central Europe before World War II, moved to the U.S. can be explained in terms of these circles. Here one finds, besides the creation of paradigms, another root of the metropolis-province division.

The mechanism does not proceed on the laboratory scale though. There are emerging economic superpowers which do not yet hold an adequate position on the scientific landscape. One of the obstacles has always been brain drain toward the old metropolises. In time of the internet, it is not even essential to drain the brains from China or India. The individual scientists may stay in Bombay or Beijing, as long as they contribute via the Web. Mind drain will suffice.

Is there a tendency toward extra-scientific criteria for the importance of some specific place in science? In times of rising dependency of academic research on third-party funds (actually: second-party funds which are not made available by a democratic state although it is responsible for all matters of national education as well as local policy and private economy direct to the evolution of science. While the basic funding for research in Germany is provided by the state and the European Fund for the development of science is considered to be only an additional source of support, the British long learned to base their financial planning primarily on resources raised in Brussels. British scientists made their own Margaret Thatcher's postulation "I want my money back!" years ago. Because of highly specific and excessively bureaucratic application procedures, scientists rely on external specialists or neglect their original academic duties and morph into application writers themselves. That way, incentive systems are driven away from the Humboldtian unity of teaching and research. Increasing neoclassical tendencies in European science policy (in the sense of "Support for success!") further strengthen metropolises and *eo ipso* reinforce provinces to stay where they are. In fact, one may even speak about a global process where (for the time being) scientific hubs in the United States set the standards and all others conform to them, i.e. subordinate.

Decisions to fund some research and not another, are decisions made by people. These decisions are, unlike prejudices, not logical consequences of pure market behavior and they are not inevitable. One may question whether it is desirable to stay in the above mentioned circle. Is it desirable for an individual scholar; is it desirable for Society? That seems all but obvious. First, many important discoveries and theories did not originate from highly competitive environments. Second, research, sometimes at least, needs time, endurance, a certain tolerance in the face of failure, maybe even relaxedness. They are difficult to achieve if the researcher has to constantly apply for the next project or contract. Dropping out, leaving the metropolis, finding a quieter place, is then an option.

Another reason for leaving well-funded research groups is the constant need for justification. For most people, for most politicians, and even for many scholars, sending a human to Mars is as useless and a waste of means as writing a commentary on Leibniz. As a result, funding goes to groups that invent patentable or at least marketable goods. Admittedly, to guess which research projects have good prospects is a hard task to perform. Should you assign large funds to somebody who promises to make gold out of lead, or to somebody who promises to build the Blue Brain, i.e. a copy of the human brain? In the first case, at least the end was serendipitous: Böttcher invented European porcelain, so perhaps there is some hope for Henry Makram and his Human Brain Project, too. The matter becomes even more pressing given that many research questions can be tackled at a very large scale only. To bring someone to Mars and back, to eventually find the Higgs bosom, etc. – all cases of "big science" – need enormous expenditures that will transform the research place into a scientific metropolis. Inversely, economic reasons may also erase a metropolis of science, as it happened in case of ITAR, the European thermonuclear facility in France.

Capitalist economics, technological needs, and finally, democratization lead to, on the one hand, still growing institutions devoted to research and technical development, and on the other hand, to the involvement of more and more people in these processes. Scientific journals provided the essential means for intellectual exchange. Important institutions connected with important journals dominated their fields and attracted people and funding. In Germany, for instance, Göttingen became a metropolis of mathematics and Berlin one of physics early in the last century. Whether one could publish results in an important journal could and did depend on where one came from, and whether one's results got acknowledged could and did depend on the journal they were published in.

As said before, province in the scientific context, be it in the institutional or in the organizational sense, is not constituted by geographical location. To be provincial does not mean to live in a province in the geographical sense, for example in Göttingen, but rather not to belong to the leading collectives. Nevertheless, Nowak's thesis that province in the sciences has nothing to do with its location is too strong. There is, besides scientific province, also commonsensical province, dark province, the place where nothing happens, wherefrom young people flee away as soon as they can. Usually, scientific institutions are not founded in such places. But the concept of province has gradations. Regarding provincial universities, one would expect them to be in a "pretty provincial" location rather than in some backwaters. Without precisely defining it, we will refer to the everyday province as geographically-social province (geosocial province, for short). Contrary to what Nowak seems to believe, there are connections between geosocial and scientific province. To begin with, the locus of cultural thought will essentially affect its inhabitants. At this stage the role model of an academic teacher also comes into play.

Scientific excellence may take a whole region out of geosocial provinciality. Cambridge, Oxford, and Palo Alto are examples. Sometimes one tries to bring out this effect by deliberate settling of ambitious academic start-ups. Money alone is not sufficient, as one may easily observe by looking at the only moderately successful new universities in the Emirates. Risks of a purely money-driven strategy are high and therefore regional policy rarely makes use of purposefully locating real or said-to-be scientific excellence. Investment alone does not create a sustainable academic culture. Normally a scientific metropolis rises with particular ease when there is metropolis in other aspects also: political, cultural, geosocial. Scientists, as all intellectually active people, are quite discerning with respect to their life environment. Top researchers are highly mobile because of their distinguished position in the scientific community. They may well live in the geosocial province as long as this is fine with their family. Inherent natural beauty of a place may positively influence the quality of scientific life. The excellent reputation of Constance University certainly does not suffer from its location at Lake Constance with the Alps as a background. Normally, very good universities are nicely situated and have a pleasant infrastructure. Scientists like to be there.

Mediocre researchers feel good at provincial universities that are situated in geosocial metropolises. Provincial universities in the province have decidedly lower esteem. From years-long empirical observation, one may derive a practical criterion out of that: academic institutions tend to be provincial if they are unable to attract top researchers and very good students, and if the faculty hurries to leave the place of employment immediately after duty. Whoever has a chance to enter another university is happy to do that. Very important persons leave and even take along some part of the institution. Usually, however, there is little chance for another position elsewhere for these scientists. They have to remain where they are regardless, even if they do not identify with the place. If nearly all of the philosophers of some university in northern Germany, all newly appointed after 1989 to replace their Eastern colleagues, need an airplane to commute to work, then their institute is provincial. If German

professors at an international university hurry to reach the earliest train to
Berlin after lectures, they transform their alma mater into a provincial
university. If the majority of the professors live far away from their
university then this is sign of scientific, as well as geosocial, provinciality.
This indication of provincialism applies to scientific provincialism of the
institution. Nowak's criterion refers to organizational provincialism. To
further calibrate the proposed measure, it needs empirical data, which are
protected by privacy restrictions. However, it seems to be a rewarding
topic for ethnographical research.

To identify provincial vs. metropolitan science, notorious rankings of
universities seem indispensable, provided, however, that data generation is
taken away from management experts and consultants and that it is handed
over to anthropologists who are experienced with the peculiarities of
academia. Who else, if not professors of a given university and those of
neighboring institutions in the national network (including more would
risk a mainly episodic picture), should be competent in judging structures
of intellectual domination within the scientific community? The task for
experts would be to clean up the avowals from emotional staining. Perhaps
a Delphi study might lead to a fairly realistic picture of the scientific
landscape, but of course, one more scientometric survey is not at the
center of interest. Reflections on scientific provinciality and of one's own
position in such a topography may seem unusual and therefore interesting.
They may help to clarify the structures of research and about conditions of
academic life.

Sometimes provinciality in a geosocial sense may be an advantage.
Provinciality as regional isolation may contribute to deeply felt familiarity
of the local elites. Barriers from neighboring areas – such as art, religion,
pseudoscience, or popular science – are lower. But so are the hurdles
between respective groups. In the province, it is almost impossible to
avoid contact with others. The level of mutual knowledge as well as
mutual animosity is often higher. Nevertheless, transgressing the
boundaries seems to be more common in the province than in the
metropolis. That is only natural: the frontiers are closer in provinces.
Compared to metropolises, the feeling of togetherness is much stronger
and may, in the form of local patriotism, lead to direct and disinterested
endowment of local scientific institutions. That way organizations may
benefit that otherwise would hardly receive adequate promotion because of
their (at best) regional importance. However, potentially tight connection
to the local sponsor may lead to even deeper provincialism – in the sense
of isolation and separation from scientific mainstream. Another positive
effect of life in the provinces is rendered possible by modern information
technology. Thanks to internet-based forms of scientific cooperation,

conference calls, maybe even distance teaching, one may afford living in scenic neighborhoods, several hundreds kilometers from the institution. Academic education certainly will suffer, and so does the spirit of the university.

The interplay between economic factors and metropolis-province-structures in the sciences is, as stated at the beginning, a dialectical one. Economic pressure on scientific institutions happens to be dangerous in several respects. If the notorious big business consultancies work out strategies on how to optimize the structure of a university, if so-called university counselors, experts from business or the financial industry, without intimate knowledge of university culture or its specific activities, actually take over the function of a supervisory board, then this profoundly changes the texture of an age-old institution. (It took a Martin Luther to successfully reorganize the Catholic Church.) The disciplinary structure of the academic organization may dissolve. As was argued in §2, rising particularization of research (and teaching) apparently leads to more academic fields and thus to new paradigms. That, however, allows the institution to announce itself a new metropolis. To repeat, the size and importance of such a metropolis is narrow. And yet that is what consultants love: a new USP[5] is established.

All this is part of an ongoing crisis of the European universities. If university education and research disperses itself then, as the self-image of faculty as a social and cultural elite vanishes, so does the significance of metropolitan science. University professors who undergo an endless ranking marathon, professors who are evaluated by first-term students, academic researchers who serve as stylish add-ons for their main sponsor's board meeting, full professors who chum up with their alumni – all that contributes to decadence of university education and undermines its role as corrective of societal development (cf. Urchs 2009). Thus, the Bologna process also prominently contributes to making academic science provincial.

4. Political Pressure

There is a whole class of actions directed at the relationship between scientific (and geosocial) provinces and metropolises, which are mainly intentional. They belong to the political area and may change the picture

---

[5] For those who are unfamiliar with consultants' jargon: 'USP' is short for 'unique selling proposition'.

considerably. For example, Jaśkowski, the Polish logician mentioned above in §2, had a professional life deeply shaped by the political influences of his time. His fate remains fascinating from our perspective after the Second World War, too. Immediately after the war, he ended up in Toruń where a university had just been established. The majority of the highly recognized professors came from Stefan Batory University in Lvov, thereby making the new university an important unit of Polish higher education. Jaśkowski had lost almost all of his manuscripts during the uprising in Warsaw and reconstructed only part of them after the war. Many of his seminal ideas, concerning e.g. intuitionistic logic, were rediscovered and published by other authors. Some of them were so exceptional that they were able to avert a similar destiny. Does that secure him the highly deserved international reputation? In no way! Published in idiosyncratic journals and in a widely exotic language, neither his papers on inconsistency tolerant calculi nor the one on causal logic found international attention in the scientific community. Within Poland Jaśkowski's work was largely known. Yet international transfer failed mainly because of the intentional inhibition of scientific contact and communication during the cold war.

A special form of geosocial province is formed by regions turned into province as a result of political processes. Occupied territories are always provincial. That has implications for scientific provinces as well. Evidently, war and revolution in part determine the status of a scientific institution. German aggression in 1939 changed even Cracow into province. Even if a researcher at Jagiellonian University had been able to escape the fatal terror and to continue doing research, the work would by no means have resulted in normal research within the worldwide scientific community. Of course, the same holds true in various degrees for all countries under the Nazi regime, even the aggressor's own country. Under the Nazi regime, German science turned provincial altogether (Wolters 2009). Some of the greatest minds of their disciplines, who managed to escape from occupied Europe, such as Albert Einstein, Alfred Tarski or Rudolf Carnap, settled elsewhere in the world and turned those places into scientific metropolises.

People behave differently depending on whether or not they come from a geosocial or scientific metropolis. As already indicated above, the whole concept of province makes sense only in relation to a center, a metropolis. To speak about metropolitan vs. provincial science presupposes established scientific communication. A metropolis without knowledge of its provinces is deficient. Similarly, the province needs to be aware of the metropolis nearby. No provincial science dwells in complete isolation. From the perspective of the metropolis, the province is a blind spot rather than a black hole – it remains quasi-invisible. The provincial peers are magnetized by the metropolis; the metropolis rarely looks back, in which direction, anyway? It is surrounded by province. Province borders on virgin soil. Metropolis prefers to look at other metropolises and has little interest in province. That is left to provinces. Therefore the topic "province in the sciences" is telling. It needs a creative mind to discover and to address it. Yet it can only be investigated by someone who became socialized in the province. Otherwise the issue would have been overlooked. That is by no means a vilification. Quite to the contrary, to become a creative thinker is certainly easier after a metropolitan education. If a student senses the reverence, which his admired academic teacher extends toward some transatlantic master-mind, it will certainly be quite hard for him to overcome that awesome two-tiered distance and to meet that colleague or his pupils at eye-level. All the greater are the merits of those who went through their provincial socialization unbroken, as Leszek Nowak did.

Typically, in the province, a true sense of style is harder to achieve. There is little chance to experience the important and the valuable. It becomes difficult to tell apart art from kitsch. For that reason province offers a preferred location to visit for third-rank representatives of metropolis. Province is the only place where they are greeted with deference. At a Central Asian university, the head of the language laboratory from Oxford University finally receives the desired respect. Inversely, top scientists often do not succeed under similar circumstances. Because real criteria for scientific excellence are not available, one is tempted to rely on outward features instead: self-confidence, urbane behavior together with well-tailored suits and expensive shoes serve as an identifying code. In a provincial university, a philosopher of repute, but with the appearance of David Lewis, was more easily overlooked than some Manager of Recruiting Service from Lewis' university. On the other hand, good teachers at provincial universities are very concerned about academic manners. Ethics of science are often highly valued, maybe because that kind of education cannot be achieved in natural workday contact with big science. Possibly everyone may report similar observations. They seem to point to interesting facets of the subject matter under consideration. Again, that might be a rewarding topic for an ethnological investigation.

Speaking about teachers opens up the topic mentioned above, education. Growing democratization of our societies enlarges the educational possibilities for many people. It is a process that first leads to a rising level of provincial intellectuality and cultural standards. Then, the comprehension of science leads to more candidates becoming intellectuals, creating a scientific metropolis. Here, modern technologies are big influences. Contrary to the situation a hundred years ago, even the inhabitants of deep geosocial provinces in industrial countries have access to comparably cheap and reliable information. Modern technology, unfortunately, also allows for new forms of exploitation. Earlier brain drains toward social metropolises aimed at pulling brilliant thinkers out of their intellectual provinces. Open societies, modern technologies, the very same means which can make life so much better, are used to perform something that might be called mind drain. People stay in their geographical and social provinces yet remain part of the scientific metropolises. Well understood, the World Wide Web even helps to cement some unjust province-metropolis structures.

So far, the political influence on the province-metropolis relationship was described rather negatively. However, based on the considerations in the previous section, one can easily see that political decisions are unavoidable in order to keep science running. Part of these decisions is directly aimed at the province-metropolis relationship. First of all, means are scarce, and it is not a wise idea to let the market alone decide how they are to be distributed (cf. Mirowski 2011). As mentioned before, from an economic perspective, there are interesting projects, important ones for the self-esteem of a nation, which need to be funded by the whole nation. A good example would be the Mars project of the United States. Whether this project is economically viable or not might be questionable. There are some projects that are simply too big to be shouldered by one institution; here CERN comes to mind. High energy research is a result of international agreements. It simply is desirable. Second, there is a relationship between socio-cultural development and the existence of scientific institutions. So, the funding of one university over another might be a decision in favor of a region.

Last but not least, there are historical and political circumstances that determine the scientific province-metropolis status of states, organizations, and scholars. One of them is the spoken language. Of course, differences due to not being a native speaker in the leading language of science can be leveled by good training. Nevertheless, writing and speaking a foreign language requires additional effort. Undoubtedly, it is a big advantage to have a unique dominating language of science. However, it also means that all books and papers not written in that language smell of provinciality.

Some people may decide not to pursue the metropolises. Such a decision may result either from a deliberate choice of provincial life or it may be due to building a metropolitan life outside the metropolises. In times of the internet, connotation of province and isolation, whether intellectual or material, seems to diminish. The anarchive of the World Wide Web may be taken as the realization of the noosphere. That phenomenon could not have been predicted by Leszek Nowak. In a specific sense, the all-available net ends any intellectual isolation and thus renders the idea of province obscure. In a sense, an intellectual province no longer prevails. The web could be turned off altogether. What would happen then? Would everything and everybody return to the previous state of affairs? Or, in a more realistic scenario, for political or financial reasons, one may exclude certain individuals or populations from using the net. Then the border would run between those in the net and the outsiders. Without access, one may be located in the province though living in the middle of Moscow, with all the well-known ambivalent implications: whoever does not waste their power and time in internet may fully devote them to science. To turn off the net, whether voluntarily or by force, means a departure into province without up-to-date news and without libraries. This, however, would imply an obvious conceptual tension: if province is characterized by the lack of originality, how may the internet (the big leveler) determine the metropolises? Would province not rather be the place for intellectual originality? This was the case at the very beginning of European science. In ancient Greece 2,500 years ago, the great thinkers started the enterprise in the provinces: in Pergamon, Milet, etc. Later the schools were moved to Athens. One has to be careful here. It is not easy to see whether these regions were provinces in the geosocial sense.

For a long time, province was exclusively and negatively connoted. If however, in times of information accessible from everywhere, province does not necessarily lower chances for intellectual development, then its meaning may change. As doubtful as it seems to acquire education through the internet, the very notion of province is gaining charming, promising overtones of deceleration, sustainability, human touch. Province today is associated with living in the scenic countryside rather than with a life without hot water to clean up with after a hard day's work in the field. One may wonder whether this will have implications for scientific province as well.

5. Pressing Virtuality

The main systematic question to be answered here is: Will the World Wide Web and other modern technologies flatten out all differences between scientific provinces and metropolises? The answer is twofold: No, there are processes that reproduce differences over and over again. But, the relationship between provinces and metropolises will change dramatically.

Theory change and especially paradigm change are responsible for a permanent occurrence of new central scientific ideas and important research. Better opportunities of communication, faster access to information will encourage the development of larger groups. Better means of collaboration should intensify research as well as competition. Therefore, more frequent changes of theories and paradigms are to be expected, which, in its turn, must dynamize the province-metropolis relationship considerably. Since the World Wide Web serves as a comparably easy searchable storage area for good ideas, fewer ideas and fewer contributions will be forgotten. Unfortunately, there will be more publications and, if the Web does not change structurally, there will be more insignificant information. To find results produced in provinces will become easier. To decide what is worth considering will be more difficult.

As we said before, home office work and even a certain amount of distance teaching via the World Wide Web makes it possible to loosen the connection between geosocial and scientific metropolises more and more. Nevertheless, markets will require their participants to feel some kind of loyalty – high quality teaching, advertising and promoting of projects as well as ideas still require bodily presence. Possibly, a special kind of Web mobility will occur: teams of researchers who collaborate in temporal groups on temporal projects. The internet supports the internationalization of all markets, hence the funding, especially of bigger projects, must become more international. European research sponsoring, as we know it, can be only the beginning.

Very important changes will occur on the political field. Already today, the World Wide Web is a means to democratize science and humanities. So, some journals are under boycott because of their price policy, books are available cost-free. Although we doubt (and hope) that high quality teaching exclusively via the net is possible at the moment, many attempts at e-learning are impressive. These are positive tendencies. Yet unfortunately, one may expect the above mentioned mind drain to accelerate. Furthermore, there will be a tendency of increasing dominance of English in the metropolitan institutions of science.

The conceptual framework of metropolises and provinces in science is, as should be clear by now, an extremely fruitful one. It incorporates numerous processes of diverse nature into one common model of reflection on philosophical, political and economic aspects of science, as well as on the role of current techniques of communication and management of scientific data. A lot more could have been said on these issues. It should be clear anyway that the approach is highly appropriate.

And yet, it is not all-embracing. It suits the discussion of problems of scientific research. But the issue of academic teaching and education is covered to a significantly lower extent. Since the subject matter of provinces in science seems to be of rising importance, there are more of interesting and open questions ahead.

EBS Universität für Wirtschaft und Recht
Chair Philosophy of Science
Department of Strategy, Organization, and Leadership
Gustav-Stresemann-Ring 3
D-65189 Wiesbaden
Germany
e-mail: max.urchs@ebs.edu

TU Dresden
Chair Philosophy of Science and Logic
Institute of Philosophy
Zellerscher Weg 31
D-01062 Dresden
Germany
e-mail: uwe.scheffler@tu-dresden.de

REFERENCES

Greenblatt, S. (2011). *The Swerve. How the World Became Modern*. New York: W.W. Norton & Co.

Fleck, L. ([1935] 1993). *Entstehung und Entwicklung einer wissenschaftlichen Tatsache. Einführung in die Lehre vom Denkstil und Denkkollektiv*. Basel: 1935. Frankfurt/M.: Suhrkamp 1993.

Kuhn, T. (1969). *The Structure of Scientific Revolutions*. Chicago: University of Chicago Press.

Kuhn, T. (1977). Objectivity, Value Judgment, and Theory Choice. In: *The Essential Tension*, pp. 320-339. Chicago: University of Chicago Press.

Mittelstraß, J. (1998). Interdisziplinarität oder Transdisziplinarität. In: *Die Häuser des Wissens*, pp. 30-42. Frankfurt/M.: Suhrkamp.

Mirowski, Ph. (2011). *Science-MartPrivatizing Science*. Cambridge, MA: Harvard University Press.

Nowak, L. ([1998] 2012). The Structure of Provincial Thought. Half Essay, Half Thesis. This volume, pp. 51-66.

Wolters, G. (2009). Philosophie im Nationalsozialismus. In: H.J. Sandkühler (ed.), *Vergessen? Verdrängt? Erinnert? Philosophie im Nationalsozialismus*, pp. 31-51, Bremen: Unesco.

Urchs, M. (2009). The Influence of Private Finance on the Scientist's Self-Image. In: P. Kawalec (ed.), *Science Management*, pp. 27-41. Lublin: Lublin University Press.

Barbara Przybylska-Czajkowska
Waldemar Czajkowski

SOME REMARKS ON THE SPACE-TIME OF CULTURE

ABSTRACT. The paper is intended as a contribution to historical materialism. The authors are not interested in any historical-philosophical, still less in exegetical, problems. They believe that this intellectual tradition is relevant for understanding the social (civilizational, cultural) transformations ongoing contemporarily and for coping with practical problems to which these transformations give rise. Among the most important changes there are those labeled as "globalization" and the rise and development of "knowledge society" and "information civilization." If one adopts this stance, one should admit that the problems of space and culture (which have been much overlooked in the Marxist tradition) deserve special attention. Discussing the ways these problems could be tackled in historical materialism, the authors use the ideas of Leszek Nowak (non-Marxian historical materialism) and Immanuel Wallerstein (world-system theory).

1. Introductory Remarks

1.1. Let us commence this text with quoting titles of two books. One of them was written by Aurelio Peccei, an Italian industrialist (top-rank *Fiat* manager), a scholar as well as the founder and the first president of the Club of Rome. The title (of the German and Polish translations; the original Italian version is differently titled) of his manifest-book is: *The Future is in Our Hands*. The second one is a work of a German philosopher, Hans Jonas. And its title is: *The Imperative of Responsibility.* These two titles express in a very brief form our general philosophical-political-ethical stance. And the content of these two books presents views which are very close to our own. Assuming this stance, one has to be confronted with the (easy and simple to be formulated, terribly hard and complex to be answered) question: what is to be done?

The answer(s) can be being looked for in a variety of ways: more or less rational. We do opt for the rational ways. And "rational" means here (at least): "knowledge-based." Thus, we need relevant knowledge. And the

In: Krzysztof Brzechczyn and Katarzyna Paprzycka (eds.), *Thinking about Provincialism in Thinking* (*Poznań Studies in the Philosophy of the Sciences and the Humanities*, vol. 100), pp. 259-287. Amsterdam/New York, NY: Rodopi, 2012.

relevant knowledge is here the knowledge of the (global) History. Moreover, the knowledge should be not only factual but, more importantly, theoretical. In other words, we do need philosophy of History or historiosophy (pace Popper's, Berlin's, Lyotard's and others' anathemas) – not in the sense of, say, Collingwood or Danto but in the old (allegedly outdated but in our view quite good) sense of Vico and Hegel, of Spengler and Jaspers, and of Karl Marx.

If one decides to contribute to the development of historiosophy, one has to choose a strategy. A few are possible. First of all, one must decide whether to start from scratch, from one's own ideas or from a body of received theories (if so, from which ones). Our decision is simple. We are going to begin our work from Marx's philosophy of History.

1.2. Now, some remarks on Marxian historiosophy. Let us make a remark of minor importance but one that is perhaps pragmatically useful. We do not want to engage ourselves in any terminological debates: should we speak of historiosophy (philosophy of History) or theory of History, or perhaps we should invent still another label for this field? – It does not matter. All these names can be regarded as (approximately, at least) equivalent. Second, we regard Marx as the single most important figure in this field but we are not prone to accept the Althusserian idea of the "epistemological break" Marx is alleged to have accomplished. Third, we opt for the methodological standards established by analytical philosophy. We do not have to declare our faith in the possibility of approaching these standards in the domain of philosophy of History. This possibility has been *constructively* demonstrated, in particular in the works of G.A. Cohen and L. Nowak.

Let us also declare that the historiosophical ideas of Marx are, in our opinion, the most success-promising "theoretical raw material," but just "raw material" (here Althusser's metaphor seems to be very convenient). In other words, we are interested neither in exegetical nor in ideological correctness. Incidentally, we think that just the collapse of the Soviet block and the related collapse of Marxism as a state doctrine have liberated Marx from these ideological fetters, which he himself, to some extent, helped to produce and to install; it is one of the great many ironies of History.

1.3. Our declaration on Marx is to be formulated more precisely now. We are not going to work on his ideas directly. On the contrary, we are going to use the results of the work already done, in particular, by Leszek Nowak and Immanuel Wallerstein.

In our opinion, these two thinkers (and their respective schools) may have produced some of the most interesting theories "made of" the Marxian "raw material." That might be a good case for starting with their works. But a more substantial reason can be given. Briefly put, Marx's philosophy of History has many lacunae, weak points, etc.[1] It so happens that some of these weak points were removed (remedied) by L. Nowak, some others by I. Wallerstein. And to put it differently, and somewhat schematically, the Marxian weak points removed by Nowak remain in Wallerstein's theory, and those removed by the latter remain in the theory of the former.

There is a serious methodological difficulty in "synthesizing" Nowak's and Wallerstein's theories. Non-Marxian historical materialism of Nowak is a very precise, systematic theory (satisfying very well the standards of, say, Carnap or of Ajdukiewicz). From the perspective of Wallerstein's epistemology and methodology (which, it should be clear, we simply reject), Nowak's historical materialism is "too abstract" ("too theoretical"). The American sociologist opts for a "middle way" (between nomothetic and idiographic approaches). For him, a "good" theory should always be an interpretation of some historical facts (processes etc. – located in "concrete" and "well defined" space and time). But, as we believe, we should not be constrained with the Wallerstein's (or, for that matter, anybody's else) own epistemological views. We can read Wallerstein's historiosophy in a different way, not determined by his own epistemology; and we are going to do so. As a consequence, our approach to the work of Nowak and to that of Wallerstein has to be different in a point: Nowak's theory has a form in which it can be "overworked" – at once; by contrast, the theory of Wallerstein has to be first "extracted" from his works and then analytically prepared to be "overworked".

1.4. Our attempt to "synthesize" Nowak's and Wallerstein's historio-sophies (or, if you wish, versions or interpretations of historical materialism) will be in various ways circumscribed. In particular, we are going to focus our attention on this domain of social reality which can be labeled with the word 'culture' (though in the Marxist tradition the concept of "social consciousness" has been more popular; the more

---

[1] Of course, the definition of them depends on one's point of view. Thus speaking about weak points of Marxian historical materialism, we mean the weaknesses as we see them. The Reader is asked to keep this reservation in mind. For the sake of "economy of place," we will not be restating this reservation at any other place at which it might be otherwise appropriate.

specific – and no doubt narrower – concept of "ideology" has been used more often).

1.5. The subject we are going to debate is so vast that it cannot be analyzed so precisely and systematically as the standards to which we adhere would demand. Thus this text is but an essay in which we will try to outline some theoretical ideas, to formulate some intuitions, and to collect some empirical (historical) observations. (Just due to this intended essayistic form of this text we have decided to resign from any references.) This approach could be validated by a metaphilosophical opinion formulated by L. Nowak: "In philosophy, breadth is more valued than precision". Or, in other words, in philosophy (at least) a "top-down" strategy seems to be more (or, at least, equally) promising than (as) "down-top" one.

2. On (the Notion of) Culture

2.1. We tend to think that neither Nowak nor Wallerstein have overcome the Marxian (Marxist) troubles with culture. This does not mean that no interesting ideas about culture were formulated by any of these thinkers. There are many of them. But they do not propose anything that could be regarded as a (historical-materialist) theory of culture.[2] We are not able to fill this lacuna. We are going but to suggest some ideas that might be useful in constructing such a theory.

2.2. As is well known, the word 'culture' has an enormously great number of definitions (for sure, the famous Kroeber's list should be enlarged today). It might be rational to avoid such terms altogether. On the other hand, it is so deeply entrenched both in ordinary and scientific-philosophical language that it might also turn out to be difficult to dispense with this term altogether. Not to mention that ambiguities may play a positive heuristic function.[3]

---

[2] To be more precise, we should rather speak of a theory of a domain of phenomena usually designated as "culture." This reservation is important since we are going to claim that one of the important results of the historical-materialist theory of culture should be the "deconstruction" of the very concept of "culture".

[3] It should be obvious that from the methodological point of view accepted in the analytical, in particular Poznań, tradition, no definition can be regarded as "proper" but only as "useful," and of course – logically correct.

A variety of intuitions are related to the term 'culture'. We start with this intuition which lurks behind this word as it has been used, for instance, in the name of the (Polish) Ministry of Culture and National Heritage. This intuition seems to be at variance with that from which various concepts relevant for, say, cultural (or social) anthropology start. 'Culture' means here a sphere (domain, branch) of human activity and of the results (products) of this activity. In slightly different (though theoretically important) terms, culture can be interpreted as a branch of (social) division of labor (and consumption), in the simplest form, as the branch of "spiritual" production, as opposed to the "material" one.[4]

This distinction has its roots reaching the beginnings of class societies millennia ago. Undoubtedly, it reflects some profound and very real social differences. The work of peasants, miners or oarsmen has been (in many respects) very different from that of priests, poets or philosophers. However, in contrast to Marx's thesis and in accordance with some ideas of Herbert Spencer and Emile Durkheim, we assume that the (social) division of labor (and consumption) has been getting more and more complex. As a result, the prospects for overcoming the difficulties with delimitating culture ("spiritual" production) and "material" production have been getting poorer and poorer.

Even if we analyze the most "elementary" (i.e. oriented at producing goods satisfying the elementary human needs) branches of "material" production (and consumption) such as the production and consumption of food or clothes, we note that also these branches are strongly interlinked with science (or with pseudo-science, cf. e.g. numerous "miraculous" diets), with advertising (visual arts), with ideologies and religions (vegetarianism, "ecological" food).

Whole "styles of consumption" or even "styles of life" should also be mentioned here. From a certain point of view, the so-called "sub-cultures" are of particular interest. And among them, "motor-bike boys" or "riders" deserve – in our opinion – special attention; this "sub-culture" is undoubtedly a product of modern industry but also of Hollywood (the movies with James Dean and Marlon Brando); the fascination with motorcycles ("Harley Davidson" may be most typical) generates strong social bonds; special sort of clothing (black leather, etc.) plays very important role, also – heavy-metal music. Some sections of this sub-culture were shifting towards criminality (perhaps of a new type "entertainment-criminality"), some others – towards some ideologies (if

---

[4] This distinction overlaps with the sociological distinction between "white-collar" and "blue-collar" workers, which had even a legal form some three decades in Poland.

shallowly adopted), for instance Hell's Angels "flirted" with Nazism.

On the other hand, the "cultural character" of some objects belonging (at the first sight) to (commonsensically conceived) culture happens to vary; let us take, for instance, music and even a very special ("utilitarian") sort of it – dance music. The "cultural character" of disco music is, in our eyes, other ("weaker") than that of Polish Mazurka, Hungarian Csardas, Argentine tango, Jewish (or Bulgarian) hora.[5]

While speaking about the difficulties with demarcation of "material" and "spiritual" labor, we have focused on the "objective" aspect of labor. Let us add that there are quite similar difficulties if we shift our attention towards this "subjective" aspect. In some respect, the labor of a surgeon is "closer" to that of a technician repairing computers than to that of an internist or psychiatrist due to the physical effort and manual dexterity it requires. And think about pilots or engine drivers. Or compare the labor of a composer and that of a musician performing her music with the respect to the physical effort required. Or note some differences between an actor playing on the theater stage and in a movie filmed in natural scenery.

Let us end these considerations with two (complementary) remarks. First, there are rather obvious differences between agriculture, coal-mining, weaving and the like, on one hand, and poetry, "pure" mathematics or metaphysics etc., on the other. Second, most types of human activities can be located somewhere in-between the "material" and "spiritual" poles.

2.3. Let us now sketch some intuitions concerning ontological (or metaphysical) status of culture, beginning with a general remark on the considerations of this kind. There is little doubt that edible wild animals and non-edible wild animals, tanks and civil air-craft, shoes and tables, religious paintings and cooking-books, lathes and cars, etc. make up a set which can hardly be regarded as "ontologically coherent" from the historiosophical (social-philosophical etc.) point of view (it seems to be much more ontologically "patchwork-like" than L. Nowak's favorite example of "vegetables"). Still, the idea that all the "material macroscopic bodies" have something in common seems to be acceptable from the point of view of, say, general ontology. The situation with "cultural objects" is more or less similar.

Let us pass now to some ontological considerations on culture. Our basic intuition could be expressed with the following thesis: Any cultural

---

[5] Perhaps this example reflects some of our idiosyncrasies. It may be so. Still, we suppose, that it does so to a degree only, and to a degree, it reflects some more objective facts.

creativity is to be understood as a form of discovery and not invention. In other words, we simply reject the concept of "invention" as opposed to that of "discovery." If the former is to be preserved at all, it is to be understood as a kind of "discovery."

As paradigmatic for our interpretation of culture, we regard the Platonist philosophy of mathematics. Of course, this is a huge subject in itself. So, only the simplest and most basic theses can be formulated here. Most briefly and somewhat figuratively put, the Pythagorean theorem had already existed (and been true) before Pythagoras (or whoever of his school) was born. Many (though surely not all) philosophers and mathematicians will agree. But how many writers, painters, composers and philosophers, will agree that, say, *Hamlet*, *Guernica* or *Tosca* had existed before written by Shakespeare, painted by Picasso, composed by Puccini? Some might say that such an idea is "obviously false." Perhaps. But we take the intellectual risk and declare that we regard this idea quite seriously and are going to develop it (if, here, in a very sketchy form).

In particular, from the point of view assumed here, all these artists (and all the others including the major and the minor ones) were just discoverers. They discovered these objects in the same (metaphysically most fundamental) sense in which Columbus discovered America (for the Old World inhabitants). We could add that electric bulb was discovered in the same sense (not invented) by Edison or Sputnik by Korolev (and his team).[6]

The very idea of, say, eternal objects of Culture (thus also of Culture as a whole) is the most fundamental, yet but the first one. The second which deserves some attention is that of "cultural space(s)." Of "musical space" composed of "all possible" musical pieces, or of "painting space" made up by "all possible" paintings. Etc. And we also assume (being aware of some, perhaps very serious, (onto)logical complications – related, for instance, to the unrestricted quantification – that may arise here) that a "mechanical space" contains "all possible" mechanical devices, a "chemical space" – all possible chemical compounds etc. Or, to apply this idea to our own domain, of "metaphysical space" i.e. the space of "all possible" metaphysical systems.[7]

---

[6] This most fundamental identity of the "essence" of discoveries of whatever sort – be they scientific, literary, technological etc. – does not exclude that discoveries in various ontological domains have some peculiarities.

[7] There is still a separate, and – in our view – important and very interesting problem that cannot be debated here. We would like to note at least that the ontology of Culture sketched here seems to have some links with Nowak's unitarian metaphysics, and has been inspired

The metaphor of space is important for at least two reasons. First, it is related to the idea of "distance" (intuitively, it is rather clear that any piano concerto by Mozart is quite "close" to any piece of this sort by Haydn, a bit more "distant" to Beethoven's concertos and very "distant" to, say, Bartok's concertos. The same can said about the paintings of Velasquez, Goya, Manet and Picasso, respectively. And this idea is "very close" (!) to that of a "geometry" (or "topology") of a space.[8]

Second, this metaphor of "space" is linked with that of "journey." History of ("mundane," "this-worldly") culture can be regarded, metaphorically speaking, as a big (and "never ending"; the question of the (im)mortality of humankind can be left aside here) journey through the space of Culture (containing various subspaces: of mechanics, of music etc.), and as a production of descriptions (or "descriptions": a "real" painting is regarded as a "description" of a sort of an, say, "ideal" painting) of the discovered areas of this space.

Is this metaphor useful? It is in our opinion. In particular, it helps to grasp clearly both the differences and the interconnections between "internal" and "external" history of any branch of culture. If we have already the geographical knowledge about America, Atlantic etc., we can explain why some journeys (taken in the given direction, in the given vessels, etc.) had more probability of success than others (in other directions, in other vessels, etc.). Which American rivers were to be discovered by European first and which later, depended on the geography of America. But the geography of America has very little to do with the fact that Spaniards decided to take the risk of such a journey at the end of 15th century, and the fact that probably no Poles even considered such an idea. Quite analogously, we might explain why, say, Special Theory of Relativity should have been discovered at the time it actually was, if not by Einstein then by another physicist, but General Theory rather could not have been discovered (at that time) by any other scientist.

2.3. Philosophy, if viewed traditionally, is not only about the World (Cosmos, Being, Whole) but also (!) about our attitudes towards the World, about our emotions it arouses, about its relations to our needs (aspirations, desires). To put it in one philosophical term, it is about values. We adhere (at least at this point) to this traditional view. Thus, let us make some axiological comments on the sketched ontology of culture.

---

by this metaphysics; what are just the objective relations between this metaphysics and the sketched ontology of culture – this question has to be remained unanswered.

[8] Let us note that from the present point of view some ideas developed by structuralism can be interpreted as confirming the "geometrical" intuitions presented here.

First, we regard knowledge as an autonomous value.[9] Thus the discovery of any (previously unknown) part of the world (which comprises in itself the "mathematical world" and that of poetry, etc.) is a value in itself. No other justification is necessary for the positive evaluation of the discovery. If this discovery has some positive side-effects for our more "prosaic" well-being, so much the better, of course.

Second, if we are to speak about the instrumental value of knowledge, we want to stress its general instrumentality broadly conceived. Such an instrumentality should be viewed (also) in relation to the rather specifically human values, and (not only) to those rooted in our biological nature, shared with all living organisms, or with all vertebrae, or all mammals. Among these values is that of freedom. Genuinely human freedom is not the freedom of choosing between strawberry and chocolate ice-cream or between a blue and a brown necktie (though one should not disrespect also this kind of freedom!), but rather it is the freedom of making the choice between, to put it rather metaphorically, living in a "Democritian" universe or in a "Pascalian" one, between living the life of a mathematician or that of a poet, inhabiting a "Kafkian" or a "Hrabalian" one. In short, the greater our knowledge of the World, the more freedom we have. And the World means here not only "our" actual (and technically manipulable) world, but also all other (possible or actual) worlds.

In the context of this axiology, we could say that one of the most central questions of social theory is: How should the social world ("our" actual world with all our books and universities, with our CDs and theaters, with governments and banks, etc.) be organized if many "journeys" in possibly various directions of the space of Culture are to be undertaken, the discoveries described and made commonly accessible (in the form of books, Internet pages, paintings and movies, etc.)?

At the end of these remarks, it might be noted that from the standpoint we accept, both the "pure" philosophy (as represented e.g. by Parmenides, Leibniz, and also the author of the unitarian metaphysics) and the "applied" and "committed" philosophy (as developed by Lukacs, Sartre, or the author of non-Marxian historical materialism) are necessary elements of "complete" philosophy.

---

[9] This does not mean that we consider the instrumental or utilitarian value of knowledge not to be important. Quite to the contrary, we take it to be very important. We only want reject some utilitarian and pragmatist stances, which do not accept any intrinsic value of knowledge.

3. Some Remarks on Nowak's and Wallerstein's Historiosophies

3.1. As we have already noted, Nowak's and Wallerstein's theories differ in many respects. In particular, the American sociologist's theory lacks virtually any anthropological foundations while that of the Polish philosopher is based on an interesting and well-developed philosophical anthropology.

Just here, let us stress very briefly two important points. First, there is no social theory without some anthropological assumptions. No society (in the strict sense, pace sociobiology etc.) can be made out of stones, ants, or even chimpanzees. In other words, some specific attributes of human beings as "elements" of society (social system or whole) must be assumed (better explicitly and not only implicitly), no matter how strongly "holistic" a social theory should be. Second, the influence one ant can exert on another one is very limited. In other words, it is the very nature of individual human being that makes it possible that society (a greater or smaller part of it) "creates" the individual. To what extent and in what ways this happens are complex and difficult questions. But that even a social theory based on an "over-socialized" conception of man is based upon a philosophical anthropology (of what sort? Sartrean or Skinnerian? Freudian or Wojtylian? – that is another important and difficult problem) is a basic methodological and epistemological fact, which can be rejected only verbally.

In view of these two remarks, Nowak's theory is better than that of Wallerstein from a methodological point of view. It is better if it is better that important assumptions of a theory be explicitly formulated rather than implicitly assumed.[10]

Let us now make some comments on Nowak's anthropology, which he calls "non-Evangelical model of man." One should not attach too much importance to this label. If one does, one can be led into interesting discussions, which are irrelevant to our topic.

The first comment is of minor importance, but it does seem to us worth making nonetheless. Nowak speaks about the rationality, the counter-rationality and the irrationality of actions. We think that this tripartite division is well-grounded (at least in its own theoretical context). Still, we have trouble with the terms used. Counter-rational and irrational (in Nowak's sense) actions seem to be rational in a wider and intuitively

---

[10] The rather sad fact that Nowak's theory has many "comparative merits" in relation to Wallerstein's theory, but is considerably less known in the contemporary social science and theory is very telling, especially in the context of problems discussed in this text.

(perhaps even more) justified sense. The idea of recognizing a situation (a set of possible actions) and, even more importantly, the idea of choice-making based upon (say) a hierarchy of values, are present in all the three types of actions. This is sufficient to justify using the term 'rational' to apply to all three types of actions.

The point here is not merely terminological. We think that Nowak's classification is far from being exhaustive. The types not contained in this classification seem to differ more profoundly from the three types contained in it than these three types differ among themselves. There are at least two kinds of actions that should be included. First, there are actions that might be characterized as manifestations of emotions (in particular, acting out of rage, fear, joy, etc.). They seem to be still awaiting theoretical conceptualization. Second, there are actions that were conceptualized (whether sufficiently well is another question) as the phenomenon imitation by Gabriel Tarde or as mob behavior by Gustav Le Bon. Obviously, these two overlap. However, they are of importance not only for group behavior as studied by social psychology but also for great social processes: revolutions, parliamentary voting, etc. It should also be stressed that the relevance of the phenomena of imitation and mob behavior (also of "emotional infection") depends on the physical concentration of people and thus upon the structure of the social space-time.

The second comment is, in our view, more important as it concerns an assumption of Nowak's anthropology, one which could hardly be called "hidden" since it is evidently easily "visible", yet it has undoubtedly not been explicitly stated and, as a result, is not discussed. This assumption could be formulated thus: any action is "locally" determined. This brief formulation has to be supplemented with some remarks on the meaning of 'locally'. This term is to be understood in two ways. First, it refers to the fact that actions are regarded in the model as, roughly speaking, "face-to-face" (or rather "person-to-person" – this reservation will be discussed a few passages below) relations. In other words, the relations between John and Peter and those between John and Paul are held to be independent. Very cursory analysis of one's everyday experience should demonstrate that this assumption has an idealizational character. We will leave the question how "strong" an idealization it is open for a while. Second, the (intended) meaning of 'locally' concerns the temporal dimension of inter-human relations. These relations are regarded as having no past, no history in Nowak's model: they are going on – "now." It is rather obvious that this temporary dimension of locality is also a manifestation of an idealizing assumption.

We think that the problem of the role played by the locality assumption is very interesting in its own right. Unfortunately, it cannot be discussed systematically here. At present, we are going to determine some links between anthropological model(s) and model(s) of social space-time. (Some theoretical comments on this notion will be made later on but we think that its intuitive understanding should be sufficient now, in the present context.)

The crucial intuitions could be presented as follows. First, the importance of interactions between various interpersonal relations of which a person is a part depends (among others) on the sheer number of these relations. In other, we would say "Simmmelian," words, the question is how our relations with other people depend on the number of these people, and in what ways? This seems to be quite difficult question. We tend to think that such a factor as "personality" (or "character," but it is also socially determined), among others, determines the mechanism of interaction between particular relations in which a given individual is involved.

And this number of interpersonal relations seems to determine the role played by the histories of these relations. According to our *ad hoc* hypothesis (probably the simplest one), if we are involved in relations with relatively few persons, these relations are rather "past-dependent." If the number of the relations is rather great, these relations tend be "past-independent." On the other hand, it is rather obvious that our inter-personal relations are not of equal importance to us. They have their "weights" in the mathematical sense of the word. It could be thought that the relations of great weight are more "past-dependent" than those of smaller weight.

Now, let us return to the problem of "face-to-face" relations. We have remarked above that it would be better to speak of "person-to-person" relations. We suppose that the "form" of such relations plays a non-negligible part in their dynamics. And by 'form', we want to understand here the real, "physical" character of the situation, in which the relation is actualized. Is it a genuine face-to-face relation (without quotation marks) or is it somehow "mediated", and if mediated then by what means: by a letter (less or more formal), by computer (e-mail), by phone, or by a third person, or perhaps by the sheer physical distance, as when parties of a violent struggle use machine guns and not swords? (Some thinkers insist that it is easier to kill someone if you do not have to be looking in the eyes of your victim.)

The dynamics (nomological structure, mechanisms of determination, etc.) of social processes is very complex. First of all, the sheer number of factors (parameters etc.) contributes to this complexity but even more important is the non-linearity (circularity, self-referentiality) of these processes. One "element" of this non-linearity deserves special attention. We have suggested that the dynamics of inter-human relations (centered "around" an individual) depends on their number and "form." They also depend upon various emotional factors (such as stress, overstimulation and the like). All these factors depend (among others) on the structure of (social) space. Whether we live rather in Heidelberg or in Frankfurt (a.M), in Princeton or in New York City, etc. determines (in a complex way) the structure and dynamics of our inter-human relations and thus – though indirectly – the way we act (rationally, irrationally, etc.).[11]

In the context of the present text (focused on the issues gathered under the umbrella of "culture"), still another aspect of anthropology should be at least sketchily discussed. All possible anthropologies can be, according to our meta-anthropological analyses, divided into four basic classes: "Sartrean" ("man is free"), "Schopenhauerian" ("man is determined by its inborn nature"), "Freudian" ("man is determined by his past") and "Althusserian" ("man is determined by society"). Each of these four classes can be subdivided further. As for the first three classes, we are not going to discuss them in any way. It is the fourth class which is the most important for the Marxian tradition. To avoid possible misunderstanding, it should be noted that in this class we can include also G.H. Mead's or W. Gombrowicz's, even M. Heidegger's anthropologies – more or less distant from this tradition. Thus some of its subdivisions will be of our interest now.

We will try to characterize just one opposition that can be defined in the "Althusserian" class. The "common denominator" of all "Althusserian" anthropologies could be formulated, somewhat simplistically, as follows: the (action of an) individual is determined by the (actions of) other people. By which people? By the Other, by members of a "reference group" or, say, by "great men" (religious and political leaders, "celebrities"). In what way(s)? Great many answers are possible. We note here but two of them: first, various "interactionist" theories (Nowak's anthropology is among them, and so is Gombrowicz's or Mead's), second, theories that can be called "indoctrinalist" (one of them seems to be assumed in Orwell's *1984*). To put it in a somewhat different way, we have, on the one hand, "Althusserian" anthropologies that regard determinative (for a man, her character, personality etc.) effects of the other's acts as side effects; on the

---

[11] Let us only point to Simmel's reflections on the influence that "great city" exerts on human emotional life.

other hand, we have anthropologies that focus our attention on the intentionality of actions performed by teachers, journalists, propagandists, etc.

Our theoretical situation may be even more complicated. It may be that the way, in which "society" influences individuals, is not "anthropologically invariant" but is changing historically. One of the simplest explanations for this changeability could draw upon the historical analysis of "material production." At the early stage of the development of material forces of production, the scope of regulating nature by people was very limited but, successively, it has been growing as the productive forces have been developing. Similarly, the means of "indoctrination," "manipulation," etc. had been for millennia and centuries rather unsophisticated. However, since the turn of the 18th and the 19th centuries, these means have been developing rapidly (popular press, photography, film, radio, TV, Internet), reaching such a level of effectiveness at the beginning of the 21st century that might be compared to the level of effectiveness of material instruments. As a result, it is plausible that the relative importance of "personality production" (of the sector of the division of labor oriented at the transformation of personality[12]) has increased to a considerable degree during these centuries.

3.2. Both thinkers, Nowak and Wallerstein, develop their conceptions on the basis of Marx's theory. Thus, it is hardly surprising that they share many ideas. One of them is a tripartite ontology of society (of social reality). According to this ontology, social reality is to be composed of economy, polity and culture. Since we are going to formulate some critical (even "destructive") remarks about this ontology, we should stress that this ontology is by no means "incidental" but does have some theoretical foundations. Most importantly, this ontology seems to be a combination of two other ontologies to be found in Marx (and Marxism). Both are dichotomous. The first one, which is more "philosophical," ("Cartesian" in a sense), assumes that social reality is composed of (social) being and of (social) consciousness. The second, which is a bit more "sociological" and more specifically Marxian (Marxist), divides the social realm into the (economic) "base" and the (political and ideological) "superstructure." It is easy to be note that if we "combine" (overlap) these two dichotomies, we obtain this trichotomy.

---

[12] The transformation is, moreover, often intended. In some cases the intention is limited (e.g. in marketing), while in other it is rather comprehensive (e.g. ideological propaganda).

Let us note that though both Nowak and Wallerstein accept this trichotomy, they differ importantly in the way they make use of it. Wallerstein is more traditional. He accepts Marxian (Marxist) economism: the primacy of economy, the instrumentality of polity and culture for economy, and their functional dependence on it. This is, in our view, his basic stance, especially, if we look at his most fundamental work *The Modern World-System*. On the other hand, in his latest and more philosophical considerations, Wallerstein emphasizes the unity and the systemic character of social reality. All the three "elements" are regarded as rather artificial constructions. (The study of the nuances of the theoretical evolution of I. Wallerstein's work must be left for another occasion.)

By contrast, Nowak makes an important step at overcoming various shortcomings of Marx's theory of history, in particular his economism. Nevertheless, he preserves Marxian "trinitarianism."[13] And this stance is, in our opinion, untenable. It should be simply rejected (and not corrected by, say, adding one or more "elements," of whatever sort). – This is a big issue, and a large separate text might be exclusively devoted to it. Thus, only a (relatively brief) list of arguments follows.

First, this tripartite division lacks a sufficiently solid theoretical grounding. However, if it were its only "weak point," we could accept it (and look for some justification for it). Unfortunately, they are more serious and hardly removable problems.

Second, various "domains" of human activity cannot be placed in one of these three "pigeonholes" in a natural, non-arbitrary way. Among them, there are some relatively (!) marginal activities, which could be, we might assume, omitted in big social models as, for instance, hooligans' activity, violence for the sake of violence without any economic, political or ideological motivations. However, there are much more important types of human practice that no social theory should leave aside. The most important among which is probably biological reproduction (or, simply put, sex). This domain crosses over virtually all the other "areas" of social life but it cannot be reduced to any of them.[14] And very close to it is the large sphere of medicine. One could with some good reasons locate it in the sphere of economy (service sector). But its close relations with science

---

[13] In fact, this is not only a Marxian view. It is a rather popular, not to say "commonsensical," element of social theory and philosophy.

[14] We should note here that non-Marxian historical materialism has some of its roots in the works of J. Burbelka and A. Klawiter. In these works, the idea of "reproductive" historical materialism was developed. It is a pity that this idea has not been developed further, which is partly due to the premature death of Burbelka.

and the idea of "physician's power over patient" (cf. M. Foucault) make it rather difficult, if possible. What about the domain of "culture"? We have already discussed this issue. No clear-cut border-line between "material" and "non-material" labor can be drawn.

Third, even in the Marxist tradition, some thinkers suggested further ramifications. One could mention here Louis Althusser who insisted on the fundamental distinction between science and ideology (thus assuming, if not explicitly, a fourfold rather than a threefold classification of practices). Somewhat similar ideas can be found in Habermas. And on the verge of this tradition, Michael Mann assumes also a fourfold division, separating politics and military activities. Some good reasons for all these fourfold divisions were given.[15]

Fourth, one could say: if not three then perhaps four or five basic domains? In our view, this is not a solution. Neither three nor thirteen nor even thirty three domains will suffice. Social ontology, on which a historical materialism is to be founded, must be fundamentally reconstructed. This is a task to be accomplished in another place. At this point, let us formulate two conditions, which any social ontology should satisfy, if it is to avoid criticism of the sort presented above. *Primo*, social reality has been and will be undergoing "structural transformations." The structure of this reality cannot be defined "once and for all." To the contrary, it should be presented as evolving and some rules or trends of this evolution should be defined. *Secundo*, the structure of social reality even statically characterized should not be thought of as composed of a number of domains. The main trouble is not with the number of domains but with the deeper underlying idea of the relevance of just one division – no matter into how many groups. Thus, the problem lies in the idea that the division is, so to speak, "one-dimensional." We need "multi-dimensional" divisions. Such divisions must take into account the fact that writing books can be a form of spending leisure-time (not very distant from playing chess or Nordic-walking) but also a form of earning a living or of propagating religious ideas. Surely, there is a "logic" of book-writing that is different from a "logic" of carpentry, and of driving a car, of giving lectures, etc. But all these "logics" basically share a character determined by the "ontological nature" of object-of-labor or object-of-consumption. If we take into consideration the "rules of game" defining a political party, a

---

[15] To give an example from other traditions, let us refer to the social ontology of Bronisław Malinowski who distinguished five sub-systems in a social system: economic, political, legal, educational and cultural. One should note the separation of politics and legal system as well as of education and culture.

Church, a corporation, etc., we will be considering something very different from these "logics." And if we are speaking about "individual strategies" of satisfying needs (of earning money, gaining social prestige, assuring physical security etc.), still another division will be described. This analysis is far from being exhaustive and systematic. Its only task has been to offer an argument for our claim that "trinitarian" social ontology cannot be just corrected but that it has to be fundamentally reconstructed.

4. Space-Time (of Culture)

4.1. The task to be accomplished in this chapter is double. First, we are going to sketch an outline of a theory of social space-time. And second, we will try to apply this theory (even in its sketchy form) to the sphere of culture.

Let us formulate the basic intuitions which, in our opinion, should be elaborated and developed into a historical-materialist theory of space-time. (It should be noted that this sketch of a theory draws mostly on Wallerstein's ideas.) We start with the general (say, ontological) intuitions. We pass to the more specific (say, sociological) in the next sub-chapter.

First (summarizing previous remarks on the anthropological assumptions of social theories), the basic elements of History (or, of social reality) are individual human actions. Even in Hegel's historiosophy, the Spirit of History is acting "through" human beings. We can add such elements to our image of History, we can see the interrelations between elements in one way or another, but – we cannot eliminate human actions. No individual actions – no History (thus, as stressed above, no historiosophy without some anthropology).

Second, we can define and analyze human actions in one way or another: as more or less "rational" or as "emotional," as "free" or as "determined," and if "determined" then "culturally," "biologically" or "socially" etc. Yet, however "humanistically" or "antinaturalistically" we define human actions, one key point is, in our opinion, beyond any (serious) doubt and debate: human actions are physical events and as such they have temporal and spatial coordinates, they are located at a "point" of physical space-time. (If you have any doubts, think about the tragedy of a blind or deaf person, or of a person lacking all her limbs.) Let us stress that this concerns even most "purely human" actions. If you are to communicate to another person some poetry or metaphysical ideas, you have to be able to emit sounds or to make some gestures, and the person has to be in the vicinity of you, or some material devices (mobile phone, computer, Internet) have to be at your hand.

Third, most of our actions need some material instruments. Be it a loaf of bread to be eaten, or a metaphysical book to be read; a pen and sheet of paper to be written with or on, or a landscape to be contemplated. Even if you do not need any particular instruments or tools while you are, for instance, speculating on mathematical or metaphysical puzzles, you need a place (be it a study or a café, a garden, or a forest path), which enables concentration. And so on.

Fourth, human beings are spatially distributed just like the (actual and potential) material instruments of their actions. These distributions can be regarded as both the most elementary and the most fundamental "layers" ("strata") of social space-time.

Fifth, social interactions are the basic type of social relations.[16] They can be viewed as physical processes (obvious reservations omitted): as transmission of information or energy, or transportation of physical objects. In other words, at the basis of social relations lies the exchange ("symmetric" or "asymmetric") of goods (including "anti-goods," such as bombs, nuclear radiation, chemical waste, etc.). Very informally speaking, space-time is a dynamic reality constituted by this perpetual movement (the second part of the term is to underscore this fact).

Sixth, an important part of this movement is the movement of human beings. To be more specific, we should speak about a great variety of this movement: starting from the its simplest forms, such as "micro-movements" at home or, say, in a factory, office etc., through "meso-movements" (holiday travels, political manifestations) to "macro-movements" such as transoceanic migrations. These movements can last for a very short time and occur very often (many movements at the working place or at home). Others can last a bit longer and occur a bit less often (daily travels to-and-from working place). Others still last up to weeks or months and are ("normally") undertaken no more than a few times during one's life-time (migrations).

Seventh, the information (energy, transportation) network that makes possible all these movements is a very important material element ("layer") of space-time. Saying this, let us note that some (!) elements of the old-fashioned historical materialism (of Kautsky, Lange and also G.A. Cohen) should be re-invoked here. As we remember, in this ("naturalist") Marxian tradition, the predominant role (that of "prime mover" of History)

---

[16] We set aside important questions that arise, in particular: in what sense are social interactions the basic type of social relations?

was ascribed to "productive forces." Quite rightly, we believe. But, unfortunately, these "forces" were understood mainly as tools (machines etc.) and their complexes. The theoretical view was focused on, so to speak, "particular" tools (from bows and spears, through ploughs and quern-stones, to looms and lathes), much less on "universal" instruments (of social reproduction) such as houses, roads, means of transportation. And just the development of these "universal" instruments has played a crucial role in the transformations of (global) space-time. And these transformations have been an essential part of the global History.

Eighth, the structure of these networks determines (in a fundamental way) the temporal dimension of space-time. In other words, the networks determine the amount of time necessary for the relocation of people and goods and for the transmission of information and energy. Obviously, the "socially relevant" distance between two "points" of the social space-time is measured in the units of time (necessary for the passing from one of them to the other). Three additional notes are to be made. *Primo*, we should distinguish between the "possible" (the shortest physically possible time) and "actual" (say, average) time. *Secundo*, various networks (road, railway, Internet) define various distances thus, we could say, various subspaces of the social space-time. *Tertio*, geographically close "points" can be socially very distant and those geographically distant can be socially rather close (it is possible that the air travel, say, from Moscow to Paris is shorter than from Moscow to a Russian village).

Ninth, the various borders (barriers etc.) constitute a very important part of the social space-time. These borders vary considerably as to their "nature." Some are "natural" in their origins: mountains, rivers, seas and oceans. But whether these geographical borders separate or connect depends on the material means of transportation. The Atlantic Ocean had been, for a very long time, a separating border. But since 1492, it has been changing to a greater and greater extent into a "connecting border." The Mediterranean Sea has, for millennia, rather connected than separated.

Tenth, the organization of space is characterized by a considerable degree of "inertia": usually a lot of energy, time, money is required to re-organize space. The larger the scale of space (of a "layer," or "level"; for instance we can speak about continents-level, regional or community one), the greater (*ceteris paribus*) the degree of "inertia". It is relatively easy to make some new streets in a town. It is much harder to build up a new city. And is extremely hard to reorganize the railway-network of a country radically, and in the case of the road-network it is probably even impossible.[17]

4.2. Having characterized social space-time in a more general (say, socio-ontological) terms, we will try now to give a more specific (say, sociological) characterization of this theoretical category. We will use some theoretical ideas of Immanuel Wallerstein.

At the very heart of his theory is located, it seems to us, the idea of, say, "asymmetry" of social space-time or, in other words, of "non-equilibrium" (of the network(s) of exchange, constitutive of the social space-time), or of unequal exchange (in terms of the "dependist" tradition), or of power, or of exploitation (in the most specifically Marxist term). This general idea takes the more specific shape of his theory of the world-system, which is composed of center (core), semi-peripheries and peripheries.[18]

In our opinion, the very idea of this differentiation of the social space deserves to be preserved. However, in Wallerstein's theory, this idea is strongly linked with his economism, which is basically taken over from Marx. As has been convincingly demonstrated by L. Nowak, economism should be rejected. If so, we should rather develop the idea of various social space-times (being subspaces of the global or universal space-time). It should be noted that we have arrived again at the problem of the division of social reality (into domains or sub-systems). Without a solution to this problem, the theory of space-time lacks an important part of its theoretical foundations. On the other hand, an intuitive or descriptive analysis of the global space-time could be helpful in solving this ontological problem. We tend to think that (some) Polish mathematical and logical departments (with Banach or Tarski) belonged to the center of world science-space (mathematics-space?) in the interwar period, while the Polish art (as represented, say, by K. Szymanowski or Witkacy) was not located in the center of world art-space. If this observation were valid, we could maintain that the very existence of the two autonomous spaces is an argument in favor of the "sociological separation" between science (mathematics and logic) and art.

---

[17] The post-war "Iron Curtain" which divided Europe in 1945-1989 might be viewed as a remnant of the Roman Empire. If this supposition is correct, it demonstrates well the permanence of, some at least, spatial structures.

[18] By the way, let us remark that the idea of "asymmetry" is a Marxian (Marxist) idea *par excellence*, while the idea of the spatial "articulation" of this "asymmetry" is an original contribution of the American thinker.

It should also be noted (a point rather missed in Wallerstein's work) that the idea of centers, semi-peripheries and peripheries, can and should be applied on different levels of space-time (or of any subspace: military, mathematical, financial etc). We can and should speak about centers of the world, of the continents, of countries and states, of regions, of cities and towns, or even of some buildings.[19] In this context, we should note that between "centers" defined for various levels there holds a *sui generis* relation of transitivity. For instance, the central cities of core countries are the central cities of the world-system and their central zones (e.g., Wall Street or London City) are the central zones of the world. But, on the other hand, it can happen that some zones of non-central cities are nevertheless central in the world. A case in point is the Vatican (assuming that Rome and Italy are not central) or Mekka (Saudi Arabia almost for sure is not world-central).

We are going to pass now to an idea which, in our view, may be the most original and interesting in the whole theoretical conception of I. Wallerstein. Putting it in a few words, we could say that he rejects the concept of "society," one which has played such an important role in social theory in general and in sociology in particular.

Let us stress (and this remark is quite important in the context of a text, which discusses Leszek Nowak's social theory, in which the concept of "society" has not only been preserved but has played a significant role) that the notion of "society" plays (though rather implicitly) also a key part in most (all?) theories in which the concept of (social) "class" is important or the most important, as in the case of Marx or Nowak. In Marx's social theory, the class is surely "something more" than a statistical concept referring to any group (any set) of slaves, landlords, capitalists, etc.

Before developing our theoretical remarks on "class" and "society" further, let us note in what way the word 'class' happens to be used in the standard theoretical and historical discourses: "American slaves," "Polish peasants," "English capitalists." It is easily seen that class is understood (in these contexts) as, so to speak, "a part of (a given) society." Thus, even in the context of the most "class-centered" social theories, the term 'society' is conceptually important: even if we want to reject any "solidarist" vision of social reality and defend a vision which regards "conflict" as central (to the social structure and dynamics), we presuppose this concept.

---

[19] To avoid some complications which could not be overcome at this place, we use here the standard geographical and legal concepts. On the other hand, the fully elaborated theory of space-time should also offer a conception of its geometry and topology.

In the case of Nowak's theory, the notion of "society" (class-divided, no doubt) is particularly important. The concepts of the "double ruling classes" and "triple ruling classes" (and, thus, also Nowakian concepts of "fashism," "totalitarianism" and "socialism") depend on it analytically.

But the theoretical meaning of the notion of "society" has not been systematically analyzed in the Marxist tradition and possibly also beyond it. This might have not been a historical-theoretical incidence. A partial explanation can be found, if we follow the analysis of the prominent Polish sociologist S. Kozyr-Kowalski. According to him, the actual sociological and theoretical content of the concept of "society" is not very distinct from that of "nation."[20] This should not be very surprising: social sciences in general, and sociology in particular, were shaped in the 19th century, in the epoch of nation-building, nation-state, nationalism, etc. Nowadays, we understand rather well that speaking of "1000 Years of the History of (this or that) Nation" is an ideological slogan, which has very little (if anything) to do with social and historical reality.[21] Thus, if we accept, as we are inclined to, Kozyr-Kowalski's thesis, the concept of "society" calls for a profound "deconstruction."

This undertaking is important since the notion of "society" (at least traditionally conceived) makes it difficult (if not impossible) to note and theoretically articulate some crucial social-theoretical problems. And just some Wallerstein's ideas can be regarded as interesting attempts directed at this goal.

Wallerstein offers a classification of "social-systems."[22] He distinguishes between "micro-systems" and "macro-systems." In spite of this perhaps somewhat misleading terminology, the essence of this distinction does not consist in the respective "smallness" or "greatness" (no matter whether understood in demographical or geographical terms). Somewhat schematically, we could say that Wallersteinian "micro-systems" are

---

[20] If you consider the way in which the word 'society' is used even in sociology, but beyond the domain of "Parsonsian" abstract theoretical reflection, you probably note that phrases like 'French society' or 'Swedish society' are amongst the most popular, whereas terms such as 'European society' or 'society of New York' seem to be much less in use.

[21] Interestingly enough, we can find quite similar ideas in E. Gellner's theory of nations and nationalisms. According to Gellner, the central idea of nationalism can be formulated very briefly and simply: the political and ethnic-cultural boundaries should overlap. And this ideological demand is to be, in his view, rooted in economic and cultural necessities of the times of industrialization.

[22] It should be stressed that the way in which Wallerstein uses the term 'system' differs considerably from the way in which it was used by Parsons. It would be an interesting theoretical exercise to analyze both, pointing out the differences as well some possible similarities.

classless while his "macro-systems" are class-based. In other words, in his "micro-systems," there is no (economic) exploitation, while exploitation is of central (systemic) importance in "macro-systems."[23]

Except for the conceptual decisions, Wallerstein does not pay any attention to the "micro-systems." He focuses it on the "macro-systems." Leaving aside his speculations about the future "socialist world government," he distinguishes two basic types of "macro-systems": empires and world-economies. Briefly put, empires are based mainly upon non-economic mechanisms (coercion, violence) of extraction of surplus value (of exploitation), while world-economies are mainly based upon economic mechanisms. In our opinion, the most original and inspiring is Wallerstein's idea of (necessary?) overlapping of "economic space" and "political space" in the case of empires (whole economy under one political rule), and of one economy associated with a system of many political entities (states) in the case of world-economies (each states controls but a part of the whole economy thus regulates to a limited degree only). The problem of the historical accuracy of this image of empires and world-economies cannot be debated here. But whatever the result of such a debate (and some critical voices seem to be justified), we are of the opinion that the very theoretical idea that lies at the bottom of this image is interesting and deserves to be developed further (in particular, to be corrected).

4.3. Let us note an interesting analogy between one of the crucial ideas of the Nowakian historical materialism and a version of the world-system theory. As regards Nowak, we think about these ideas on which his conception of totalitarianism, fascism and socialism are based. These ideas can be summarized in two points. First, analytically, three different types of class domination (division) should be distinguished: economic (owners/laborers), political (rulers/citizens), spiritual (priests/believers). The distinguished ruling classes can overlap.

They can overlap to a degree though Nowak himself discusses but the "extreme" cases: the single classes (no overlapping at all), the double classes (e.g. owners-rulers) and the triple class (owners-rulers-priests) where total overlapping occurs. It can be supposed that the tendency toward monopolization of all sources of social power is among the most fundamental and universal for all "Nowakian societies."

---

[23] It is quite possible that there are some, historically changing, relations between the geographical (demographical) magnitude and the level of exploitation.

As regards the world-system theory, to make it compatible with Nowak's theory, we should distinguish three (sub-)spaces: economic, political and cultural.[24] For each of this spaces, we can distinguish its center, semi-periphery and periphery. A very natural (and, we think, important) questions arise. First, how strong will be the tendencies to the "convergence" of these centers? Second, what are the consequences of separation (or of overlapping) of these centers?

It should also be noted that we could link this analysis with the idea that centers can de distinguished at various levels of the world space-time: from the center of the global system to the centers of cities, towns etc. For instance, some interesting phenomena can be observed at the level of "countries." In France, for instance, Paris can be regarded as the center of power, economy and culture. A more or less similar (perhaps even more strongly articulated) phenomenon is characteristic of Hungary. The space of Great Britain seems to be a bit more differentiated: at least in the domain of culture Cambridge and Oxford play a role comparable to that of London. The centers in Germany had been still more dispersed before the unification: the political capital in Bonn, the economic (financial) in Frankfurt (a.M.), the cultural in Heidelberg, Goettingen. When studying the micro-spaces of towns and cities, we should find analogous differentiation.

4.5. Let us now turn to some remarks on spatial dimension of culture. At this point, we will collect some observations to be accounted for by a historical-materialist theory.

Let us start with repeating an observation already made, though in another context. We have noted that Polish mathematics and logic gained an important position in the 1918-1939 period, probably justifying the claim that it belonged to the center of a domain of the world culture. If this claim is true, it could be developed in two directions. First, a nation ("country") can become the world center of a special, narrowly defined, domain. Second, if the criteria of the quality of work are more precisely defined then the "monopolistic power" of a country (countries) is weaker than in domains with less clear criteria.

A somewhat similar phenomenon can be registered in the case of the history of art. We think here of painting at the turn of 19th and 20th centuries. It is not only Paris which was at that time the "world capital of

---

[24] We are critical of this "trinitarian" ontology. Nevertheless, introducing a more adequate but also more complex ontology would make the present reasoning less clear. So, we will assume the simplified "trinitarian" ontology for the presentation of the key idea.

painting." And we could be more precise, and speak first about Montmartre, then about Montparnasse. (We doubt that the role of New York City nowadays is similar to that of Paris.) As regards literature, philosophy and the humanities, especially if we take into account some currents (existentialism, structuralism, Western Marxism) again not only Paris, but a part of it, so-called *Rive Gauche* (The Left Bank) played the crucial role. We could go even further and note the role played by various *cafés*, in particular those located at the Boulevard Saint-Germain (*Café de Flore* etc.).[25] It should be noted that *cafés* had taken (to a degree) the role played by the aristocratic *salons* of the 18th and 19th century. In this context, we should also mention the institution of *colleges* (Cambridge, Oxford; UK), which spatially link the professional (scientific and didactic) activity with the private lives of the scientists.

Having made some remarks about Paris, we would like to say a few words about another city, Budapest. At the turn of the 19th and the 20th centuries, Hungary was by no means a part of economic or political center, even if one regards it as a part of Franz Joseph's monarchy (which then experienced its Indian summer). However, astonishingly many persons who greatly contributed to the world culture were Budapestian in origin. Among those born around 1900 in the capital of Hungary, you can find such persons as J. von Neumann, A. Szent-Gyorgi, B. Bartok, A. Kocstlei, G. Polya, M. Polanyi, K. Polanyi, G. Lukacs, K. Mannheim (this list could be easily enlarged). It is noteworthy that quite a few of the listed and many others important persons were Jews. They lived on the "borderland" of three cultures (in the ethnic sense of the word): German, Hungarian and Jewish. Perhaps some demands of such a life (such as mastering two or three languages from early childhood) stimulated intellectual development. On the other hand, they were all born in Budapest but only Lukacs was active there for a longer time.

From cities, let us pass now to regions or, to be more precise, to one very important region, California. It could be regarded as the center of computer industry (though some reservations are necessary: Microsoft has its headquarters in the state Washington; but Apple is there). California is also the center of another domain existing on the border area of traditionally defined economy and culture: film industry. That these "cultural" industries have been developing in California can hardly be regarded as historical "accident." On the other hand, the fact that one of these industries has its main location in Los Angeles, and the other – in

---

[25] In fact, the role played by Lvov's *Scottish Café* is well known in the development of the Polish mathematics between the World Wars.

San Francisco (the center of counter-cultural movements at a time) seems also to have some quite deep social roots.

4.6. As noted at the sub-chapter 3.1, one of the characteristic (onto)logical attributes of social reality are its non-linearity, self-referentiality and reflexivity. Moreover, we think that each of these attributes can assume some (higher or lower) degrees or levels. And, as already noted, we tend to agree with Spencer and Durkheim that social (but also biological) evolution can be characterized, in particular, by the increase of the respective degrees or levels.

In this sub-chapter, we are going to make some comments on one aspect of this trend: we want to say something about the relation between space-production and culture.

Let us first stress that we have deliberately used the phrase 'space-production'. Like many other objects of social reality, so also the social space (in the literal sense, i.e. understood as composed of roads and streets, of houses and factories, etc.) has been to a lesser and lesser degree a side-effect of other activities, and to a greater and greater degree a result of intentional activities. Some elements of urban planning existed already in the Ancient Greece, not to say Rome. In the 20th century, there were more or less successful attempts to plan spatial deployment of industry in some states. There are also some elements of planning on the scale of a whole continent as in the EU transport projects.

In our times, it is technically and economically possible to transport a car-factory (or even steel-works) from one continent to another. Yet neither today nor in any foreseeable future will we be able to "transport" even an average town, not to mention a metropolis of million dwellers. The "inertia" of the social space has decreased in some respects but it has increased in others. Taking all the tendencies together, we believe that the total "inertia" has increased. Thus decisions concerning the structure of (global, regional, etc.) space are among those of most long-lasting effects for global History.

As we have stressed in this text, the structure of the (physical/geographical) space is of great importance for the social reality. It affects virtually all types of social relations: from the most intimate interpersonal relations to the international relations of various sorts.[26] Thus

---

[26] Let us but note the instances of large-scale planning that would deserve an analysis of this sort. In the first instance, undoubtedly, the creation *ab novo* of the new capital of Brazil, Brasilia, would deserve to be considered. Also of great interest is the experience of Auroville, which is a town whose creation was inspired by philosophical ideas.

the spatial planning is a form of social planning. We should add that it is an instance of social planning of considerable scale and that planning is unavoidable (pace Popper, Hayek etc.). Thus what we need today is not a criticism of the very idea of (social) planning (as formulated most prominently by these two thinkers). What we need today is a "critique of (social) planning reason."

5. Final Remarks

5.1. Let us commence these end remarks with some considerations on the "policy of philosophy." First, this phrase is but a shortcut (similar to that of "policy of Poland"). Actually, we think here about the policy of, say, "world community of philosophers." And here a substantial question arises: Does such a community exist at all? The answer is, in our opinion, rather negative (generally speaking, though some reservations and qualifications should be added here). And if so, we should ask other questions: Should (could) such community exist? Let us assume that the answer is positive. The problem of how to construct (establish, organize) such a community arises. There is no simple solution to the problem. Rather a whole set of undertakings (as to congresses and conferences, as to book-series and journals, professors and students exchange, etc.) should be thought of.

Now, the second point: an interpretation of the role of philosophy. Philosophers differ not only on metaphysical, ethical, etc. issues; they differ profoundly in the understanding of the "nature" ("essence") of their own activity, of philosophy. In spite of these (metaphilosophical) differences, some intuitions seem to be quite widely accepted. In particular, the intuitions centered around such ideas/concepts like "whole," "totality," etc. Consequently, philosophy can be interpreted as "having to do" with the "whole of culture," with the relation between science and art, natural and social sciences, ideology and religion, ethics and law. We are not going to claim that such an interpretation of the "nature" of philosophy is in any sense most "proper" or "correct". But we tend to think that so understood philosophy could (and should) play a very important social (cultural) part.

We could formulate this opinion on the social role of philosophy somewhat differently. Very briefly put, one could say that philosophy is (should be) a critique of culture. Of its content, of the way of its production and consumption, philosophy (when playing this role) should offer ideas for the transformation of education, research, media as well as relations between culture, politics and business.

Ending these considerations, let us declare that we are aware of the fact that we are undertaking a subject present in the modern (intellectual) history since, at least, mid-18th century - since the time of Enlightenment, of *les philosophes*, of Encyclopedists, through that of Russian and Polish *inteligentsia* and Western intellectuals, to the participants of the cultural (ideological) part of the Cold War. This history must be a source of often bitter, disappointing and skeptical reflections. From our point of view, postmodernism (no doubt a very complex phenomenon, one of which but one aspect is here referred to) can be interpreted as a manifestation of these reflections. Being aware (as we hope we are) of all the dangers of political commitment, nevertheless we do opt for it. We think that the spectrum of possible attitudes of intellectuals is much richer than the alternative of a Benda's *"clerc"* and Lenin's "party intellectual." And a detailed analysis of the last 250 year history should teach us how to protect oneself against the dangers resulting from linking politics and culture too eagerly and too simplistically.

5.2. In the previous point we have opted for political commitment of philosophers in particular and of intellectuals in general. The general reasons for this option are outlined at the very beginning of this text. Now, let us try to formulate some goals philosophers should try to attain. Very briefly put, we could speak about one goal: the democratization of global culture. Of course, this is a very sketchy formulation and some, even if concise, comments are needed. Perhaps, it would be convenient to formulate some reasons why democratic character of global culture should be an important goal.

First, the democratization of culture means (all other factors being equal) stronger pluralism, greater diversity of culture. And pluralism (diversity) is in the interest of all in the long run, including those that profit from monopolization and cultural homogenization in the short run. The basic arguments for this claim are evolutionary (in the broad sense). They refer to the link between diversity (of, say, "internal states" of a system) and adaptability (of this system) to changing conditions of its reproduction.

Second, the democratization of culture seems to be linked with the idea of dignity of each human being, her (intellectual) freedom and spiritual autonomy. In other words, democracy means equality. And since we opt for radical democracy, we must opt in particular for equal access to public debates, to defining values, goals, norms.[27]

Third, culture is an important instrument for reproducing group-identity. And though we think that global cultural identity (human kind as social/cultural community) is desirable, we also think that preservation of smaller (in particular, national) identities is, from our point of view, an important value. Thus, a great and difficult question (and a practical challenge) arises: How to construct the institutional and material infrastructure of world culture so that the global and the local are possibly well balanced.[28]

5.3. One of the most popular slogans of ecological movements (coined allegedly in 1972 by the French-American biologist Rene Dubos) says: "Think globally, act locally!". This phrase can be interpreted not only "geographically," but also "sociologically". The word 'global' refers to the whole social system and 'local' to a subsystem (to a branch of human practice). As professional philosophers, we tried to answer the question "What we, philosophers, could do as philosophers?". It might be that the answer(s) we have suggested will not be accepted. We want to believe that at least the way we formulated the question will be accepted. It is founded on a simple moral rule: you should demand from yourself at least as much as you demand from your neighbors.

Silesian University of Technology
Katedra Stosowanych Nauk Społecznych
ul. Roosevelta 26-28
41-800 Zabrze
Poland
e-mail: barbaraprzybylska@wp.pl
      waldemarczajkowski@wp.pl

---

[27] We are aware of all "logistic" and "technical" difficulties. The ideas formulated here should not be regarded as practical demands but rather as criteria with which the social-cultural transformations should be evaluated.

[28] Of many special questions to be solved, one should be mentioned in the context of this paper. The location of great many headquarters of various international organizations reflects the material-cultural dominance of the USA and Western Europe.

Katarzyna Paprzycka

THE INTELLECTUAL SUPERPOWER

AN ATTEMPT AT A CORRECTION OF NOWAK'S MODEL
OF PROVINCIALISM

ABSTRACT. The paper has two goals. First, I reconstruct Nowak's model of provincialism. Second, I argue that there is no room in this model for an intellectual superpower. An intellectual superpower is not simply a paradigm that is buttressed by economic, political, and social resources. Rather it is a network of paradigms that recognize one another as such. In so doing, they create a new quality on the scientific arena that surpasses all single paradigms.

Nowak offers a general characterization of a provincial thinker as someone who is "incapable of venturing [her] own judgment" ([1998] 2012b, p. 54), who is fully dependent on the center. In particular, she accepts the judgments of the center without questioning them and, furthermore, takes it upon herself to spread the word onto even more provincial others. Nowak tries to conceptualize the phenomenon of provincialism in science. He suggests that the phenomenon should be understood at a social level in terms of forceful cognitive domination of some scientific communities by others.

First, Nowak distinguishes three basic types of research: creative, correctional and applicational. All three types of research are essential for the development of science. The primary function of a scientific school is to assemble researchers, who carry out correctional and applicational research, to further develop the creative research carried out by the founder of the school (§1.1). Second, Nowak claims that while progress in science involves the cooperation of all kinds of research, creative research is particularly valuable because it is rare and yet crucial if science is to truly accomplish its theoretical mission (§1.2). Third, he considers the social conditions that are optimal for the development of science and casts

In: Krzysztof Brzechczyn and Katarzyna Paprzycka (eds.), *Thinking about Provincialism in Thinking (Poznań Studies in the Philosophy of the Sciences and the Humanities*, vol. 100), pp. 289-301. Amsterdam/New York, NY: Rodopi, 2012.

them in terms of several desiderata. I show that one can think of two desiderata as fundamental in the sense of comprising Nowak's more specific desiderata. These two fundamental desiderata are the desideratum of pluralism and the desideratum of respectful coexistence (§1.3). Fourth, he argues that the phenomenon of provincialism is a form of forceful cognitive domination of whole scientific communities by other communities (§1.4). This suggests that the roots of provincialism are the inequalities between the communities – some have social, economic, and political power that others lack (§1.5). The resulting model (§1.6) carries with it certain practical implications (§1.7). In §2, I show that the model does not take into account what I call an intellectual superpower. Its addition requires a modification of Nowak's model.

1. Nowak's Model of Provincialism

1.1. *On the Necessity of Scientific Schools*

Using the apparatus of the idealizational conception of science (Nowak 1971; 1980),[1] Nowak distinguishes three types of research: creative, correctional, and applicational. Applicational research involves the acceptance of a given theory, in particular of all the factors that it identifies as relevant ("essential" in Nowak's terminology). The researcher's own input consists in finding new yet unsolved problems, to which the theory can be applied as is, i.e. without introducing any changes to it. Someone who engages in correctional research accepts the core of the theory (in Nowak's terms: she accepts the principal factors and the idealizational laws) but introduces changes in the way the laws are concretized, most commonly by taking into account new secondary factors. The most original type of research, creative research, involves changes in the factors taken to be principal and so changes in the idealizational laws. A creative researcher proposes an essentially new way of thinking about the phenomena, proposes a new paradigm, to use Kuhn's term. Moreover, such a proposal invites further development – it invites research of a correctional and applicational type.

The development of a paradigm is usually carried out within a scientific school, which encompasses the founders of the school (the authors of the paradigm), correctors as well as applicators. While it is

---

[1] While Nowak offers the distinction of these three types of research in terms of his own idealizational theory of science, the distinction can be accepted more generally.

possible that individual persons can fill more than one role (someone equipped with a complete research personality can do creative, correctional as well as applicational research), it is often the case that individual scientists fill only one of those roles. Nowak takes scientific schools to be necessary for a complete development of a paradigm. Even if the founder of the school is also capable of engaging in correctional and applicational research, there is just so much that he can do during his lifetime.

1.2. *Theoretical Progress*

What underlies Nowak's account is a broadly Popperian rather than a positivist vision of progress in science (see e.g. Nowakowa and Nowak 2000, esp. chapters 4 and 26). We should not think of science as aiming merely at the theoretical accommodation of all facts. The goal of science is not merely to develop a theory that will catch the most facts into its explanatory net. Rather, the aim of science is also to search through the space of possible theoretical nets with view not only to catching the most facts but also with view to how deep an understanding of the facts (and of the previous theories) a theory can provide.

Creative research has thus a special value in science for it enables the exploration of the space of possible theoretical approaches. Correctional and applicational research is also necessary to complete the creative vision and make it a contender in the space of scientific theories.

1.3. *Optimal Conditions of Theoretical Progress*

Nowak proposes certain desiderata, which could be construed as capturing the ways, in which science should be done or in which science would be done under ideal conditions. Nowak's desiderata (cf. [1998] 2012a, pp. 39-42) can be seen as addressing two main concerns: the optimal development of science and the optimal social relations among researchers that would enable such a development. They can be captured by two overarching desiderata: the desideratum of pluralism and the desideratum of respectful coexistence.

The above vision of scientific progress straightforwardly leads to the *desideratum of pluralism*, according to which there should be as many different paradigms as possible. This is the only way in which the space of possible theories can be properly explored. The fulfillment of this desideratum requires the fulfillment of Nowak's desideratum of idealization (there should be as many different ways of idealizing phenomena as possible) but also of the desideratum of concretization (the initial propositions need to be developed to approach empirical reality) or of the desideratum of criticism (ruthless confrontation with empirical reality is necessary for an adequate development of a paradigm). The satisfaction of the desideratum of pluralism presupposes also that the desiderata of correspondence and of dialectical refutation are satisfied (the paradigm developed ought to encompass the theoretical accomplishments of older theories).

The second overarching desideratum concerns the optimal relations among the adherents of the paradigms. The *desideratum of respectful coexistence* requires the adherents of a paradigm not only to tolerate the supporters of other paradigms (desideratum of tolerance), not only to recognize the right of sufficiently competent researchers to engage in any type of research (desideratum of egalitarianism) especially creative research (desideratum of freedom of research), but also to do everything possible to offer constructive criticisms (desiderata of criticism, of clarity, of interpretational charity, of co-thinking) and to actively engage in attempts to save paradigms that lose adherents (desiderata of support for the weakest and of co-thinking).

1.4. *Provincialism as Forceful Cognitive Domination at Community Level*

Nowak understands provincialism as a form of forceful domination. He presents the idea of cognitive domination thus:

> *x dominates y cognitively* if *y* grants *x* the right to creativity while *y* grants himself only the right to correct or apply *x*'s ideas (considered to be creative). It follows that someone who is dominated adopts the role of an applicator or corrector of ideas he takes to be dominating. ([1998] 2012b, p. 62)

Cognitive domination may be of a forceful nature but it need not be such. Sometimes the division of labor within a scientific school, for example, takes place naturally, i.e. it is the result of the natural predispositions of the participants. A scientist might be simply either only capable or only interested in doing, say, applicational research. Such a person is dominated cognitively by the author of the theory he is applying. It should be stressed, however, that such domination need not even have any social dimension. For the scientist might have no contact with the author of the theory. In fact, she might long be dead.

It happens not infrequently, however, that cognitive domination is accompanied by social domination. Though Nowak does not say so explicitly in ([1998] 2012b), it is clear that this can happen within a scientific school as well. The founder of the school might use social force to ensure her cognitive domination over other researchers (if only by offering support, employment, etc., only to those who fill the roles she prepares for them).

It is important to recognize that the founder of the school is able to exert such force only if she has enough social power, afforded by her position in the scientific community. The position, moreover, has many dimensions, among them: political (e.g. participation in scientific decision-making bodies), economic (e.g. funding), and social (e.g. recognition by other scientists). A young researcher with a brilliant idea for a new paradigm, who does not have any position within the wide scientific community, would not (yet) be capable of exerting this kind of social domination over the potential adherents of his school. He can be the agent of cognitive domination over others but not of forceful cognitive domination.

According to Nowak, provincialism is a form of forceful cognitive domination over whole scientific communities.

> [The effect of provincialism] consists in the fact that there appear whole "master" communities, who take themselves to have and who are taken to have leading roles, while the role of all the other communities is downgraded to a supplementary role – of "correctors" community or "applicators" community. In other words, the provincialism effect consists in the replication of the natural structure of a cognitive community at a more global inter-community level, whether in between continents, in between countries, or in between academic centers. ([1998] 2012b, p. 63)

One could ask the question whether a spontaneous, natural, and unforced, cognitive domination of one master community over another is possible. This would have to mean that the members of a community spontaneously do only correctional or applicational work. This can well be imagined when we consider a temporal cross-section of a community. It might so happen that all biologists working now in city x do only correctional or applicational research. The matter changes dramatically, when we consider the community of biologists in city x without restricting attention to a specific time. It is extremely hard to imagine that it would be impossible for there to appear biologists in city x with creative abilities. To the extent that we are concerned with the actual phenomena of cognitive domination of one community over others, we can safely ignore spontaneous and natural cognitive domination and concern ourselves exclusively with forceful cognitive domination.

1.5. *The Root of Provincialism*

We saw that in the case of a scientific school, the founder of the school can exercise forceful cognitive domination over the other members only if he has enough power (political, economic, and social). Likewise, forceful cognitive domination of one "master" community over other communities presupposes that the "master" community has a sufficiently powerful position within the wider scientific community. The precondition of exerting force by the master community is its appropriate political, economic, and social standing. In other words, a precondition of the exercise of forceful cognitive domination is ultimately an inequitable distribution of resources. This seems to be the root of provincialism on Nowak's account. It is because the resources are not distributed equitably among scientific communities that the forceful cognitive domination of some communities by others is made possible. It should be stressed that the resources in question are not only economic. There are numerous factors at stake such as prestige, tradition, political participation, and so on, and so forth.

If this reconstruction is correct then, on this model, provincialism (understood as forceful cognitive domination by one community over others) would be erased if the distribution of resources among scientific communities were equitable. This would mean, for example, that especially the economic and political resources ought to be distributed among the supporters of different paradigms, paying particular attention to paradigms on the verge of social extinction (desideratum of support for the weakest) and to new paradigms (support for young and fresh ideas). On the other hand, the support for a hitherto strongly supported paradigm that has attracted a lot of supporters (at the expense of others), which ceases to develop (few corrections or applications are available), should be lessened. And so on. Such a distribution of resources would provide a political, economic and – with time – also social infrastructure, that would be capable of supporting the desideratum of pluralism as well as the desideratum of respectful coexistence.

1.6. *The Model of Provincialism*

Let us try to summarize what we have been saying on behalf of Nowak. The identification of the root of provincialism in a preexisting power structure leads one to accept the following claim about an idealized situation, viz. where the distribution of resources is equitable:

(1) If the distribution of resources among scientific schools were equitable then the desiderata of respectful coexistence and of pluralism would be satisfied.

If the development of new and unexplored paradigms were encouraged (economically, politically, and socially), while the further exploration of overexploited and progressively more fruitless paradigms (degenerative research programs, to use Lakatos' term) were discouraged, one could expect a relatively optimal realization of both desiderata: the ideal of paradigm pluralism as well as the ideal of the optimal social relations among scientists.

We know very well that the distribution of resources is not equitable. It is not equitable in at least two ways. First, certain scientific communities are favored on geographical or social grounds. This leads to the domination by some well-endowed and powerful communities over others. In this sense, e.g. the top American universities dominate most of the universities in other parts of the world.[2]

Second, however, certain scientific schools of thought (communities in a scientific sense, which may but need not be identifiable in geographical terms) can be favored. The history of Artificial Intelligence research, for example, has witnessed a well-nigh extinction of the neural network (perceptron) research paradigm in the 1960s. The successes of the classical approach resulted in an inequitable distribution of resources, which brought neural network research to a virtual standstill. Nowadays, we are witnessing a surge of neuro-enthusiasm It is justified to the extent that the new methods of brain imaging have opened vast research areas, which rightfully attract researchers. We are, however, also slowly reaching the point where the neuro-fashion is hurtful to other paradigms of research in psychology, for instance (see also Paprzycka 2010).

If the distribution of resources among scientific communities is skewed in the direction of certain geographical communities then the effect of geographical provincialism in science will take place. Certain (geographical) communities will dominate others. It will be expected that only the members of those powerful communities produce new ideas while the members of the remaining communities will be expected to apply them or, at best, to correct them. A new idea proposed by a member of the provincial community will not be recognized as worth pursuing. Frequently, it will simply not be recognized (for members of master

---

[2] Incidentally, when we talk of the leading role of American universities, we in fact ignore the majority of American universities. There are over four thousand universities and colleges in the United States, of which only a couple have the leading role that we allege "American universities" to have. Most of the other schools are just as, or even more, provincial than European provincial universities. This great polarization of American education (unmatched in Europe) has a lot to do with a much greater social mobility of Americans.

communities do not read provincial work[3]). And even if it is recognized, it will not be treated charitably. It bears emphasizing too, as does Nowak, that "it is easy to come up with genuine counterarguments against [any] new proposal" ([1976] 2012, p. 71).

If the distribution of resources is skewed in the direction of certain paradigms then the effect of intellectual provincialism will take place. Certain schools of thought will dominate others. Such domination involves the lack of recognition of another school as developing a paradigm in its own right.

If the above reconstruction is correct, then Nowak's model (with some extensions) can be summarized by means of two claims:

(1) If the distribution of resources among scientific schools were equitable then the desiderata of respectful coexistence and of pluralism would be satisfied.

(2) If the distribution of resources is skewed either to favor certain geographical communities or to favor certain schools of thought then science will be witness to geographical or intellectual provincialism, respectively.

This model locates the roots of provincialism in the preexisting power structure. One way to counteract the phenomenon would be to change the distribution of resources. This is, however, hardly something that an individual can do. The question one may ask is: What *can* an individual scientist do?

1.7. *"What Should I Do?"*

This question leads us to a certain tension in Nowak's account ([1998] 2012b). The model of provincialism we have reconstructed describes the phenomenon at a social level. The root of the problem is a preexisting power structure, which is not something that we can individually hope to do anything about. Yet, Nowak also insists that provincialism properly speaking is to be located at the level of an individual.

> Where is the province? The definition has nothing to do with geography: province is where one thinks not on one's own account but on account of another. Provincialism consists in a certain type of academic career (especially in the humanities). It consists in that one does not have the courage to challenge the world but that one makes oneself into a carpet, on

---

[3] It is also hard to blame them for that, as they have lots to read anyway. In fact, one can see provincialism (alongside specialization, for example) as a way for scientists to deal with too great a traffic of ideas.

which the "world" can march to the Vistula or the Amu-Darya. Provincialism is not defined by one's address but by the type of one's mentality. Who thinks on her own account, who is the center for herself, is not provincial – even if she lives in the countryside in Góra Kalwaria or Vitebsk. ([1998] 2012b, p. 65)

In other words, each and every one of us has the power as an individual to rise up to the structure of geographical or intellectual domination and not accept the role of a provincial thinker that we have been assigned. This will be extremely hard to do (see also Nowak [1976] 2012). It will require a lot of self-determination and non-conformism. We better do it when we are young. But, yes, we can do it.

However, if the model sketched above is correct and the geographical or intellectual domination is in place, then such a heroic attitude on the part of an individual will not change the social structures. As the author of non-Marxian historical materialism (e.g. Nowak 1991), Nowak is of course fully aware that this is so. Even if the individual scientist creates new paradigms, which will be completed by his followers, they will tend to be downplayed. On the other hand, if the paradigm is really successful, it becomes more and more difficult for it to be ignored or downplayed. Moreover, the well-developing paradigm might acquire some new resources, which might strengthen it further.

The prescription, which Nowak addresses especially to the young researchers ([1976] 2012) is to have the courage to think in novel ways, to depart from well-trodden paths, to develop their own paradigms, to create their own schools of thought. To wit, we, the provincial thinkers, have nothing to lose but our own chains.

The picture is not overly optimistic. The provincial thinker who has lost her chains may only count on obstacles in her path. Still, Nowak's picture is optimistic in the following sense. It is part and parcel of the situation that were it not for the uneven distribution of resources, the provincial thinker and her paradigm would play the same role as the mainstream thinker and his paradigm. This gives the direction for development. The provincial thinker should try to amass as many supporters (perhaps from other provincial thinkers), as many resources, etc. to match the situation of the mainstream thinker. There is a possibility then that, given the right support, she could reach the position of the mainstream thinker and that, in the end, she will be recognized as an equal partner.

(2*) If a (once provincial) school of thought acquires the same resources (political, economic, and social) as the mainstream school of thought, they will be equal to each other.

As a result, the relationship of domination will cease and so will the relation of provincialism: members of the once provincial school of thought will no longer be treated as such.

In what follows, I claim that this picture is overly optimistic after all. It ignores the possibility of an intellectual superpower.

2. An Intellectual Superpower

On Nowak's model, the basic unit of the wider scientific community is that of a scientific school. A scientific school is built around the founder(s) of a paradigm and develops the thought of the master(s). An intellectual superpower can be understood as a social community of different scientific schools, who recognize one another but do not recognize other scientific schools on the same rights (either do not recognize them at all or treat them as provincial thinkers). It exhibits all the symptoms of an "old boys' network," which is inaccessible to outsiders.

An intellectual superpower is a community that can create the illusion for its members that real science is done only within its bounds. In fact, the postulates of pluralism and of respectful coexistence are, by definition, as it were, fulfilled within the bounds of the intellectual superpower. The superpower comprises multiple paradigms, multiple schools, which are recognized, criticized, supported, and discussed within the intellectual superpower.

The top American universities mentioned earlier are a good approximation to an intellectual superpower.

The position of an intellectual superpower vis à vis external schools of thought is fundamentally different from the relation we have been considering in §1 between the well-endowed and powerful mainstream school of thought and poorly endowed provincial schools of thought. In the latter case, there is a possibility (even if often infeasible) that the schools could become equal partners. In the case, of an intellectual superpower, on the one hand, and an external school of thought, on the other, there is in principle no such possibility.

The external school of thought could, on certain (very improbable) conditions, become an equal partner with the schools of thought that comprise the intellectual superpower. In such a case, the external school would simply be absorbed by the intellectual superpower. At least two conditions would have to be satisfied. First, it would have to recognize the schools of thought that comprise the superpower and, what is harder to achieve, it would have to be recognized by them. Second, the process of intellectual recognition would have to be accompanied by an appropriate distribution of resources. A school of thought that is incapable of matching the others in infrastructure will not be treated as an equal partner.

The second condition seems practically impossible to fulfill for most provincial schools. But it is even harder to meet the first condition. For the intellectual world of the superpower becomes closed off and impermeable to outsiders. Because it comprises many different paradigms of thinking, it tends to produce the illusion among its members that all possibilities are covered and that any idea worth the while will find its place on the intellectual ground of the superpower. Any external ideas that do not tie with the cross-paradigmatic accomplishments of the superpower will tend to be rejected as groundless, as not worth pursuing, or as not formulated sufficiently precisely, clearly, or intelligibly. They will appear as if formulated in a different language. The meaning of the proposals will be unclear to the members of the superpower until the proposed ideas are tied in appropriate ways to the ideas that have surfaced in various areas of the intellectual superpower.[4] (So much for independence!)

What of the possibility of founding an alternative intellectual superpower? One could perhaps venture the hope – albeit with admittedly even smaller chances of success than in the above case – that with the right sort of effort, one could develop another alternative intellectual superpower capable of competing with the given one. A lot of effort would be required. Not only would one have to develop one's own paradigm sufficiently, not only would one have to find the right sort of resources, but one would also have to develop a sufficiently strong network of coalition partners (other provincial paradigms). Could such an alternative, once provincial, superpower not match the mainstream superpower under the right sort of conditions?

I believe the answer to be negative. The ideas formulated within the bounds of the alternative superpower will be as unintelligible to the members of the mainstream superpower as the ideas of an external school of thought. To the extent that they can become intelligible, they will need

---

[4] The social and psychological processes at work are very complex and worth investigating. Let me offer an anecdotal illustration. As a graduate student at one of the top American philosophy departments, I had a conversation with another graduate student from an equally prestigious department at one of the conferences. She described to me her project in ethics. It was clear to me that she was grappling for some ways of handling idealization. I suggested to her the work by L. Nowak and by N. Cartwright. When I met her at another conference, she was extremely grateful for having pointed her to Cartwright's work as it unblocked her progress on the project. When I asked what she thought of Nowak, she did not remember my having mentioned him.

to be absorbed and tied to the preexisting ideas, i.e. assimilated. But the two intellectual superpowers will not be seen as equal. They will not be seen as addressing the same issues. Rather, the alternative intellectual superpower will be seen as engaging in a different discipline. (To some extent, we can observe such a situation on the example of analytical and continental philosophy.)

If so, however, then some of Nowak's prescriptions for the provincial thinker can be repeated: not to accept the role of a provincial thinker, to try to propose new ideas, to charter new territories, to develop new paradigms, to try to amass as many supporters and resources as possible. Yet, they should be supplemented with the admonition that one should not do so in abstraction from the paradigms that comprise the intellectual superpower. Without the recognition of what has been done by the paradigms that comprise the intellectual superpower, one will never – not even under the best political, economic, and social conditions – be recognized by them.

Conclusion

In sum, the proposed correction of Nowak's model can be captured thus:

(1′) If the distribution of resources among scientific schools were equitable and there were no intellectual superpower then the desiderata of respectful coexistence and of pluralism would be satisfied.

(2′) If the distribution of resources among schools of thought are equitable but there exists an intellectual superpower then the desiderata of respectful coexistence and of pluralism are not satisfied for science at large but rather only within the bounds of the intellectual superpower. The external schools of thought occupy provincial roles even if they dispose of the same resources as any individual school of thought within the intellectual superpower.

(3a) If there exists an intellectual superpower and the distribution of resources is skewed to favor (for whatever reasons, geographic or intellectual) some of the schools of thought within the intellectual superpower, the phenomenon of secondary provincialism occurs – the desiderata of respectful coexistence might be violated even within the intellectual superpower. Under some conditions, this might even lead to the exclusion of one of the schools of thought from the intellectual superpower.

(3b) If there exists an intellectual superpower and the distribution of resources is skewed to favor (for whatever reasons, geographic or intellectual) one of the schools of thought external to the intellectual superpower, then while the position of that school of thought is strengthened, it will not affect the existing relation between the intellectual superpower and that school of thought until and unless mutual ties of recognition are established (in which case the once provincial school of thought can be absorbed into the intellectual superpower).

It is easy to see that the relation between (1′) and Nowak's (1) is the relation of dialectical correspondence. The proposed model is a correction of Nowak's model in his sense of the term.

University of Warsaw
Institute of Philosophy
Krakowskie Przedmieście 3
00-927 Warszawa
Poland
e-mail: kpaprzycka@uw.edu.pl

REFERENCES

Nowak, L. (1971). *U podstaw marksowskiej metodologii nauk.* Warszawa: PWN.

Nowak, L. (1980). *The Structure of Idealization.* Dordrecht/Boston: Reidel.

Nowak, L. (1991). *Power and Civil Society. Toward a Dynamic Theory of Real Socialism.* New York: Greenwood.

Nowak, L. ([1976] 2012). Models of Scientific Research. This volume, pp. 67-74.

Nowak, L. ([1998] 2012a). On the Hidden Unity of Social and Natural Sciences. This volume, pp. 15-50.

Nowak, L. ([1998] 2012b). The Structure of Provincial Thought. Half Essay, Half Thesis. This volume, pp. 51-66.

Nowakowa, I. and L. Nowak (2000). *Idealization X: The Richness of Idealization (Poznań Studies in the Philosophy of the Sciences and the Humanities,* vol. 69). Amsterdam/Atlanta, GA: Rodopi.

Paprzycka, K. (2010). Is Neurobiology of Personality Inevitable. A Philosophical Perspective. In: T. Maruszewski, M. Fajkowska, M.W. Eysenck (eds.), *Personality from Biological, Cognitive, and Social Perspectives,* pp. 13-27. Clinton Corners, NY: Eliot Werner Publications.

Symposium

Canadian Journal of Continental Philosophy
Revue canadienne de philosophie continentale

Editor: Antonio Calcagno, Ph.D.
With an International Editorial Board
www.c-scp.org

Symposium publishes articles, interviews, and book reviews within the various traditions of Continental European thought, including existential philosophy, phenomenology, philosophical hermeneutics, critical theory, de-construction, poststructuralism, and postmodernism. The journal publishes two issues per year and is peer-reviewed.

Some recent articles:
ALAIN BADIOU, "Comments on Simon Critchley's Infinitely Demanding"
UGO PERONE, "Public Space and its Metaphors"
BERNARD STIEGLER, "Un entretien avec Bernard Stiegler"
VITTORIO HÖSLE, "The European Union and the USA"
DAVID B. ALLISON, "Who is Zarathustra's Nietzsche?"
GRAEME NICHOLSON, "Justifying Your Nation"
JOSEPH MARGOLIS, "Toward a Phenomenology of Painting"
CALVIN O. SCHRAG, "On the Ethics of the Gift"
JOHN CAPUTO, "Good Will and the Hermeneutics of Friendship"
CLAUDE PICHÉ, "Habermas et la mission de l'université contemporaine"
DAVID CARR, "Rereading the History of Subjectivity"
RICHARD KEARNEY, "Postnationalism and Postmodernity"
BABETTE BABICH, "Reading David Allison's Reading the New Nietzsche"
ALPHONSO LINGIS, "Divine Illusions"
ALAIN BEAULIEU, "A Conversation with Charles Taylor"

Annual individual subscription (includes membership in CSCP): $60.00 /
Library subscription: $95.00

To subscribe, please send payment by check (payable to the Canadian Society for Continental Philosophy) to: Dr. Antonio Calcagno, Editor, **Symposium**, Department of Philosophy, King's College at the University of Western Ontario, 266 Epworth Avenue, London, ON N6A 2M3 Canada. Please address any queries to acalcagn@uwo.ca.

THE JOURNAL OF THE BRITISH SOCIETY FOR PHENOMENOLOGY

An International Review of Philosophy and the Human Sciences

EDITOR: DR ULLRICH HAASE

Volume 43, No.1, January 2012
Foucault: Politics, Power, Pleasure

The *JBSP* publishes papers on phenomenology and existential philosophy as well as contributions from other fields of philosophy. Papers from workers in the Humanities and Human Sciences interested in the philosophy of their subject will be welcome. All papers and books for review to be sent to the Editor: Dr Ullrich Haase, Department of Politics and Philosophy, The Manchester Metropolitan University, Manchester M15 6LL, England. Subscription and advertisement enquiries to be sent to the publishers: Jackson Publishing and Distribution, 3 Gibsons Road, Heaton Moor, Stockport, Cheshire, SK4 4JX, England.

THE JOURNAL OF THE BRITISH SOCIETY FOR PHENOMENOLOGY

An International Review of Philosophy and the Human Sciences

EDITOR: DR ULLRICH HAASE

Volume 43, No.3, October 2012
Silence, Language, World

The *JBSP* publishes papers on phenomenology and existential philosophy as well as contributions from other fields of philosophy. Papers from workers in the Humanities and Human Sciences interested in the philosophy of their subject will be welcome. All papers and books for review to be sent to the Editor: Dr Ullrich Haase, Department of Politics and Philosophy, The Manchester Metropolitan University, Manchester M15 6LL, England. Subscription and advertisement enquiries to be sent to the publishers: Jackson Publishing and Distribution, 3 Gibsons Road, Heaton Moor, Stockport, Cheshire, SK4 4JX, England.

CONSTRUCTIVIST
FOUNDATIONS

Constructivist Foundations is an international peer-reviewed academic e-journal. The journal publishes original scholarly works in all areas of constructivist approaches that support the idea that cognitive structures are actively built by one's mind rather than passively acquired.

It recognizes the fact that constructivist approaches vary in how much influence they attribute to constructions and which domain (rational-linguistic or biological-bodily) they are considered to populate.

These approaches include: radical constructivism, second-order cybernetics, biology of cognition, theory of autopoietic systems, enactive cognitive science, first-person research, neurophenomenology, consciousness studies, and non-dualism, among others.

Constructivist Foundations appeals to philosophers, psychologists, scholars of educational research, cultural studies and communication sciences, and researchers working in artificial intelligence and cognitive science.

It is listed in, among others, the *Science Citation Index, The Philosopher's Index* & the *European Reference Index for the Humanities.*

The journal appears three times a year and is free to its currently more than 5000 subscribers. It is open to unsolicited contributions from authors and also produces thematic special issues, which are published in print, too. Previous special issues focused on constructivism and society, the conscious self, non-dualizing philosophy, radical constructivism, and mathematical constructivism.

} "The world consists of what matters"

• • • • • • • • • • • • • • • • •

For more information, please see its web page at

http://www.univie.ac.at/constructivism/journal

Nation – Gesellschaft – Individuum

Fichtes politische Theorie der Identität

Herausgegeben von Christoph Binkelmann

rodopi

Orders@rodopi.nl—www.rodopi.nl

Amsterdam/New York, NY
2012. X, 369 pp.
(Fichte-Studien 40)
Paper €80,-/US$104,-
E-Book €72,-/US$94,-
ISBN: 978-90-420-3567-6
ISBN: 978-94-012-0842-0

USA/Canada:
248 East 44th Street, 2nd floor,
New York, NY 10017, USA.
Call Toll-free (US only): T: 1-800-225-3998
F: 1-800-853-3881

All other countries:
Tijnmuiden 7, 1046 AK Amsterdam, The Netherlands
Tel. +31-20-611 48 21 Fax +31-20-447 29 79
Please note that the exchange rate is subject to fluctuations

Bolzano & Kant

Edited by Johannes L. Brandl,
Marian David, Maria E. Reicher
and Leopold Stubenberg
Guest Editor: Sandra Lapointe

Amsterdam/New York, NY
2012. IV, 353 pp.
(Grazer Philosophische
Studien 85)
Paper €75,-/US$98,-
E-Book €68,-/US$88,-
ISBN: 978-94-012-0833-8
ISBN: 978-90-420-3558-4

USA/Canada:
248 East 44th Street, 2nd floor,
New York, NY 10017, USA.
Call Toll-free (US only): T: 1-800-225-3998
                        F: 1-800-853-3881
All other countries:
Tijnmuiden 7, 1046 AK Amsterdam, The Netherlands
Tel. +31-20-611 48 21 Fax +31-20-447 29 79
Please note that the exchange rate is subject to fluctuations

Orders@rodopi.nl—www.rodopi.nl

Giordano Bruno

An Introduction

Paul Richard Blum
Translated from the German by Peter Henneveld

Giordano Bruno (1548–1600) was a philosopher in his own right. However, he was famous through the centuries due to his execution as a heretic. His pronouncements against teachings of the Catholic Church, his defence of the cosmology of Nicholas Copernicus, and his provocative personality, all this made him a paradigmatic figure of modernity. Bruno's way of philosophizing is not looking for outright solutions but rather for the depth of the problems; he knows his predecessors and their strategies as well as their weaknesses, which he exposes satirically. This introduction helps to identify the original thought of Bruno who proudly said about himself: "Philosophy is my profession!" His major achievements concern the creativity of the human mind studied through the theory of memory, the infinity of the world, and the discovery of atomism for modernity. He never held a permanent office within or without the academic world. Therefore, the way of thinking of this "Knight Errant of Philosophy" will be presented along the stations of his journey through Western Europe.

Amsterdam/New York, NY
2012. XI, 128 pp.
(Value Inquiry Book
Series 254)
Paper €30,-/US$39,-
E-Book €27,-/US$35,-
ISBN: 978-90-420-3555-3
ISBN: 978-94-012-0829-1

USA/Canada:
248 East 44th Street, 2nd floor,
New York, NY 10017, USA.
Call Toll-free (US only): T: 1-800-225-3998
F: 1-800-853-3881

All other countries:
Tijnmuiden 7, 1046 AK Amsterdam, The Netherlands
Tel. +31-20-611 48 21 Fax +31-20-447 29 79
Please note that the exchange rate is subject to fluctuations